T0226110

Springer Water

The book series Springer Water comprises a broad portfolio of multi- and interdisciplinary scientific books, aiming at researchers, students, and everyone interested in water-related science. The series includes peer-reviewed monographs, edited volumes, textbooks, and conference proceedings. Its volumes combine all kinds of water-related research areas, such as: the movement, distribution and quality of freshwater; water resources; the quality and pollution of water and its influence on health; the water industry including drinking water, wastewater, and desalination services and technologies; water history; as well as water management and the governmental, political, developmental, and ethical aspects of water.

More information about this series at http://www.springer.com/series/13419

Wolfgang Kinzelbach · Haijing Wang · Yu Li ·
Lu Wang · Ning Li

Groundwater Overexploitation in the North China Plain: A Path to Sustainability

Springer

Wolfgang Kinzelbach
Institute of Environmental Engineering
ETH Zürich
Zürich, Switzerland

Yu Li
Beijing Normal University
Beijing, China

Ning Li
Institute of Environmental Engineering
ETH Zürich
Zürich, Switzerland

Haijing Wang
Hydrosolutions Ltd.
Zürich, Switzerland

Lu Wang
Institute of Environmental Engineering
ETH Zürich
Zürich, Switzerland

ISSN 2364-6934 ISSN 2364-8198 (electronic)
Springer Water
ISBN 978-981-16-5845-7 ISBN 978-981-16-5843-3 (eBook)
https://doi.org/10.1007/978-981-16-5843-3

This Springer imprint is published by the registered company Springer Nature Singapore Pte Ltd.
The registered company address is: 152 Beach Road, #21-01/04 Gateway East, Singapore 189721, Singapore

With contributions by

Yuanyuan Li, Jie Hou, Lili Yu, Fei Chen, Yan Yang (Sects. 2.1.7, 2.2, 2.3, 4.2.1 and 5.4)
General Institute of Water Resources and Hydropower Planning and Design (GIWP), Beijing, China

Haitao Li, Wenpeng Li, Kai Zhao (Sects. 1.3, 4.1, 4.2.1, and 4.3.2)
China Institute of Geo-Environmental Monitoring (CIGEM), Beijing, China

Jinxia Wang (Sects. 2.1, 2.2 and 3.2)
China Center of Agricultural Policy, Peking University, Beijing, China

Tianhe Sun (Sects. 2.1, 2.2 and 3.2)
Hebei University of Economics and Business, Shijiazhuang, China

Yanjun Shen (Sects. 3.1 and 5.3)
Centre for Agricultural Resources Research, Chinese Academy of Sciences, Shijiazhuang, China

Jianmei Luo (Sect. 3.1)
Hebei GEO University, Shijiazhuang, China

Silvan Ragettli (Sect. 4.2.4 and Appendix A.2), Beatrice Marti (Boxes 4.2 and 4.3, and Appendix A.7)

Hydrosolutions Ltd., Zürich Switzerland

Foreword

The agricultural sector increasingly relies on groundwater abstraction for irrigation in many regions of the world. Expansion of irrigated agriculture and groundwater use has contributed to increased food production and has improved food security. At the same time, the depletion of groundwater resources by excessive pumping has become a common challenge in key agricultural production areas across the world, threatening the sustainability of production. Over-pumping of aquifers causes ecological and environmental degradation of vegetation, wetlands, and streams and reduces the ability of aquifers to serve as a buffer for weather extremes induced by climate change. Unsustainable use of groundwater needs to be urgently addressed through sustainable rehabilitation and management strategies, in order to assure the availability of groundwater, today and for future generations.

Recognizing the groundwater over-exploitation issue in the North China Plain, the Swiss Agency for Development and Cooperation and the Ministry of Water Resources of the People's Republic of China co-launched the project "Rehabilitation and management strategies of over-pumped aquifers in a changing climate" in 2014, in partnership with the China Geological Survey. The leading project implementation partner on the Chinese side was the General Institute of Water Resources and Hydropower Planning Design (GIWP), while the leading project implementation partner on the Swiss side was the Swiss Federal Institute of Technology (ETH) Zürich.

The overall goal of the project was to test and implement groundwater management and water saving policies in order to strengthen the capacity for adaptation to climate variability and climate change. Any management strategy must be based on reliable data. A main element of the project was thus to establish a real-time groundwater monitoring and control system. This monitoring data, coupled with state-of-the-art modelling, serves as a basis for the development and implementation of different policy instruments in the fields of water resources and agriculture.

The present book features the impressive results of the second phase of the project. The project developed, tested, and implemented an array of innovative approaches and tools for monitoring and modelling groundwater abstraction, as well for supporting sustainable management decisions. To name just a few: For the first time in China, groundwater abstraction has been monitored on such a large-scale using electricity consumption as proxy indicator. Cutting-edge groundwater models, both real-time and data-driven, were established and a web-based comprehensive decision support system (DSS) was developed. The DSS entails tools for water quota planning, scenario analysis, estimation of actual crop water demand, remote sensing based automatic mapping of cropped areas, and a data base, all integrated in one platform. The project also developed a groundwater game, which mimics the agricultural practice in the North China Plain and was used in a serious game approach to assess farmers' reactions to policy changes.

With all the monitoring, modelling, and decision support in place, SDC will continue the dialogue with sub-national and national governments in China to support the uptake of the findings in existing and new policies for sustainable groundwater management. Many of the findings of the project and all the tools developed under it are also highly relevant for other areas in the world suffering from groundwater depletion. We are thus convinced that the findings and tools presented in this book can provide inspiration and advice in support of our common goal of managing groundwater more sustainably across the globe.

Felix Fellmann
Head of International Cooperation
Division
Swiss Embassy in the People's
Republic of China
Beijing, China

Acknowledgements

First and foremost, we would like to thank the Swiss Agency for Development and Cooperation (SDC) for the courage of taking on a multi-year commitment in supporting and financing this Sino-Swiss cooperation project "Rehabilitation and Management Strategies of Over-Pumped Aquifers under a Changing Climate". Without SDC's firm commitment to support this project, the knowledge we share in this book would not be there. Special thanks should be given to SDC staff Dr. Liyan Wang, Dr. Manfred Kaufmann, Dr. Felix Fellmann, Dr. Jacqueline Schmid, and Dr. Philip Zahner whose help and support throughout the project have been greatly appreciated by the project expert team. Dr. Liyan Wang shaped the initial design of the project. She was personally involved in the implementation process together with the project team and attended all major workshops and numerous field visits. Without her enthusiastic involvement, the project could not have been implemented as smoothly as it was. We would also like to thank Dr. Manfred Kaufmann and Dr. Liyan Wang for their valuable comments, which helped us to improve the manuscript and make it more accessible for the general reader.

We owe great gratitude to Mr. Andreas Goetz, the deputy director general of the Federal Office of Environment in Switzerland at the time, who grasped the importance of sustainable management of over-pumped aquifers through innovative technology. Mr. Goetz communicated between the Ministry of Water Resources and the Swiss Agency for Development and Cooperation to initiate a dialogue on the project. Without his strong support and help, this project would not have come into existence.

We would like to thank Ms. Nongdi Wu and her colleagues from the Ministry of Water Resources in China, who persuaded the Swiss expert team to pay special attention to the groundwater management problem of the North China Plain, which became the main focus of the project. Special thanks should be given to Ms. Jing Xu and Mr. Ge Li who actively supported us in their role as members of the project steering committee. We acknowledge all the local partners in the Departments of Water Resources in Hebei Province, Handan Prefecture, and Guantao County. Without their diligent work in implementing project ideas and their great support, our field activities, including more than 300 pumping tests, an aquifer recharge experiment and more than 20 field visits for stakeholder consultation, trouble shooting, game tests and surveys, could not have been accomplished. We owe great gratitude to Dr. Junfang Gu and Mr. Hongliang Liu, whose support from the beginning has helped us to choose Guantao County as the pilot site. They and their colleagues, Mr. Xiantong Yan and Ms. Xinmei Chen, accompanied the expert team during almost every field visit to Guantao.

Special gratitude is due to the local implementation team in Guantao County. Mr. Huaixian Yao has been our most valuable local expert, whose local knowledge and great experience in dealing with local problems have been a highlight for our team. Mr. Huaitao Wu, Mr. Guangchao Li, Mr. Dongchao Yang, Mr. Depeng Fan, and Mr. Fei Gao all contributed to the tailor-made decision support system in the project with their knowledge and experience. We would also like to acknowledge Mr. Fei Li from the State Grid Hebei Electric Power Co. Ltd., staff from Guantao Electric Power Supply Company, and the village electricians of Shoushansi Electric Power Supply Agency, who were willing to share their knowledge regarding both electricity metering and farmers' pumping habits. We also owe our gratitude to Guantao Department of Agriculture and Rural Affairs for sharing information on crop planting area and local irrigation practice.

We would like to acknowledge the extraordinary support provided by the China Geological Survey (CGS) by conducting a detailed hydrogeological survey of Guantao and surroundings, by drilling new groundwater observation boreholes, both along the border of Guantao County and inside, and by delivering observation data monthly on time and without fail according to the project needs. The groundwater level observation points on the border of Guantao made the innovative groundwater modelling of an administrative unit as well as the separation of inside and outside contributions to drawdown possible. Special thanks should go to Mr. Yuan Zhang from China Institute of Geo-Environment Monitoring (CIGEM) for his reliable and diligent work of sending monthly groundwater data until the end of the project.

Special thanks should be given to Prof. Shuqian Wang, whose knowledge and experience, accumulated in his work in water resources management of the region over the past decades, have been enlightening to the project team. Professor Wang generously allowed his students at Hebei University of Engineering to participate and help in the pumping tests, game tests, and game survey. His valuable input has been a treasure for our work in Guantao and Handan.

We would like to acknowledge the game designers from Zürich University of Arts (ZHdK), Mr. Réne Bauer and Mr. Livio Lunin, who co-designed and implemented both board game and online game "SavetheWater", which provided an excellent tool to test farmers' feedback to policies. We would also like to thank our colleagues Prof. Pan He (Southwest University of Finance and Economics, Chengdu), Mr. Jakob Steiner (from hydrosolutions Ltd.), Mr. Dominik Jäger (Geoprävent Ltd.), Mr. Gianni Pedrazzini (ETH Zürich), Mr. Alexandre Mérrilat (ETH Zürich), and Mr. Andreas Hagmann (ETH Zürich), who have been involved in our earlier field activities, and whose work has become an integral part of this project. We further acknowledge the assistance in climate change projections for Guantao by Dr. Nadav Peleg (ETH Zürich).

We owe our deepest gratitude to Prof. Paolo Burlando, who has taken the financial responsibility and co-lead role for the project within ETH. Without his generous help, the project could not have been continued after the opening phase.

Finally, we would like to thank the farmers in Guantao, who participated in our consultation meetings, game tests, and surveys and watched with interest the pumping tests, metre installation, and drone flights for remote sensing validation. They have always been willing to share their views in terms of crop planting and groundwater use. They and the farmers in North China Plain in general should be given credit for fulfilling the task of feeding a huge population under severe resource constraints. We acknowledge their willingness to save water and thank them for cooperating with us during the project.

Contents

Abbreviations and Units

Abbreviations

ADB	Asian Development Bank
BTH	Beijing-Tianjin-Hebei region
CART	Classification and Regression Tree Algorithm
CAS	Chinese Academy of Science
CCAP	China Centre of Agricultural Policy, Beijing University
CGS	China Geological Survey
CIGEM	China Institute of Geo-Environmental Monitoring, Beijing
CMIP	Coupled Model Intercomparison Project
CMIP5	5th phase of Coupled Model Intercomparison Project
CNCG	China National Administration of Coal Geology
CNY	Chinese Yuan
CORDEX	Coordinated Regional Climate Downscaling Experiment
DSS	Decision Support System
DWR	Department of Water Resources
EnKF	Ensemble Kalman Filter Method
EPSA	Electric Power Supply Agency
EPSC	Electric Power Supply Company
ET	Evapotranspiration
ETH	Swiss Federal Institute of Technology, Zürich.
GCM	Global Circulation Model
GEF	Global Environmental Facility
GHG	Greenhouse Gas
GIWP	General Institute of Water Resources and Hydropower Planning and Design, Beijing
GRACE FO	Gravity Recovery and Climate Experiment Follow-Up
GRACE	Gravity Recovery and Climate Experiment
GTDSS	Guantao Decision Support System
GUI	Graphical User Interface

ICT	Information and Communication Technology
ID	Irrigation district
IPCC	Intergovernmental Panel on Climate Change
IWHR	China Institute of Water Resources and Hydropower Research, Beijing
MAR	Managed Aquifer Recharge
MEA	Millennium Ecosystem Assessment
MNR	Ministry of Natural Resources, People's Republic of China
MWR	Ministry of Water Resources, People's Republic of China
NASA	United States National Aeronautics and Space Administration
NCP	North China Plain
NILM	Non-intrusive Load Monitoring
RCM	Regional Circulation Model
RCP	Representative Concentration Pathway
SDC	Swiss Agency for Development and Cooperation
SJZIAM	Centre for Agricultural Resources Research, Shijiazhuang, Chinese Academy of Sciences
SLF	Seasonal Land Fallowing
SLFP	Seasonal Land Fallowing Program
SNWT	South-North Water Transfer
StW	"Save the Water" Game
SYB	Statistical Yearbook
TDS	Total Dissolved Solids
UI	User Interface
USGS	United States Geological Survey
WUA	Water User Association
WUE	Water Use Efficiency

Units

a	Year
mu	Chinese Measure of Land Area, 1 mu = 1/15 ha
CNY	Chinese Currency (Yuan Renminbi), 1 CNY ≈ 0.14 Swiss Franks
d	Day
yr	Year
ha	Hectare
Mio.	Million
Bio.	Billion
t	Ton
°C	Degree Centigrade

Chapter 1
Introduction

The depletion of aquifers by excessive pumping is one of the prominent global sustainability issues in the field of water resources. It is mainly caused by the water needs of irrigated agriculture. The North China Plain is a global hotspot of groundwater overexploitation. Since the 1980s, groundwater levels dropped by about 1 m/year mainly due to the intensification of agricultural production by a double cropping system of winter wheat and summer maize. The consequences of declining groundwater levels include the drying up of streams and wetlands, soil subsidence, seawater intrusion at the coast and rising cost of pumping. The depletion of storage makes the production system more vulnerable with respect to climatic extremes associated with climate change. Sustainable management of aquifers keeps groundwater levels between an upper and a lower red line. While the upper red line is designed to prevent soli salinization, the lower red line is motivated by ecological requirements, water quality constraints or infrastructural concerns. Global water balances are useful in identifying the scope of the problem and the size of efforts required to restore a sustainable pumping regime. Adequate local action needs a local analysis. Guantao County is selected for such a local analysis in the North China Plain.

1.1 Groundwater Over-Pumping and Consequences

Aquifer depletion caused by excessive pumping has been described in the literature over the last two decades (Foster and Chilton 2003; Kinzelbach et al. 2003; Konikow and Kendy 2005), with the Ogallala aquifer in the United States' Midwest being the first and iconic case of a large, significantly depleted aquifer (Wines 2013). The Millennium Ecosystem Assessment Report (MEA 2005) identifies more hotspots of aquifer depletion due to agricultural irrigation, including California, Spain, Pakistan, India, and the main case of interest here, the North China Plain (NCP) (Fig. 1.1). In all these cases, aquifer depletion shows through constantly declining groundwater

© The Author(s) 2022
W. Kinzelbach et al., *Groundwater Overexploitation in the North China Plain: A path to Sustainability*, Springer Water,
https://doi.org/10.1007/978-981-16-5843-3_1

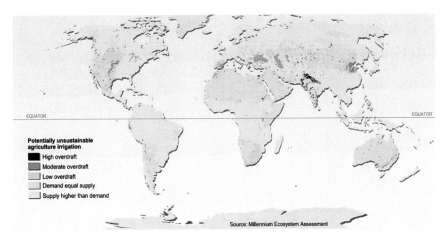

Fig. 1.1 Hotspots of unsustainable agricultural groundwater abstraction, indicating North China Plain, Northwest India, High Plains Aquifer and Southern California in the USA as moderate to high overdraft regions. *Source* MEA (2005)

levels. In the North China Plain, for example, rates of decline of up to 2 m per year since the 1980s have been observed.

A global vision of large-scale groundwater depletion has been provided by the Grace mission, a satellite mission, which has been measuring the gravity field of the earth and its changes in time since 2002 (http://www2.csr.utexas.edu/grace/ove rview.html).

Temporal changes of the gravitational field are caused by displacement of large water masses over time. The water losses on the landmass not only include water from the melting of glaciers but also from the depletion of aquifers. Comparison of Figs. 1.2 to 1.1 shows the coincidence of high agricultural groundwater abstraction and mass-losses (reddish tones). A thorough analysis of areas of water loss and water gain and their respective causes is given in (Rodell et al. 2018). Figure 1.3 shows the mass change in a rectangle covering the North China Plain (between 35° and 40° N and 114° and 117.5° E) using the mascon approach. The gravitational signal has been decreasing, indicating a loss of 39 cm of water column (or about 2.4 cm/year) between 2004 and 2020. The associated area is about 165,000 km^2.

This leads to a total mass loss of 64 km^3 or an average annual loss of 4 km^3/year. An analysis by Feng et al. (2013) taking into account an area of 370,000 km^2, identifies a storage loss of 8.3 ± 1.1 km^3/year (2.2 ± 0.3 cm/year) from 2003 to 2010. Figure 1.3 shows that there is quite some variation in the decline rate, and one must be careful when comparing figures for different periods in time. Further, the attribution of the signal to an area introduces uncertainty. Ascribing the total loss of 4 km^3/year to the unconfined aquifer and assuming a specific yield of 0.03, it translates to a groundwater level decline rate of about 0.8 m/year on average.

Why do we bother about aquifer depletion? Aquifer depletion inherently contains a mechanism by which abstraction will decrease with a declining groundwater table.

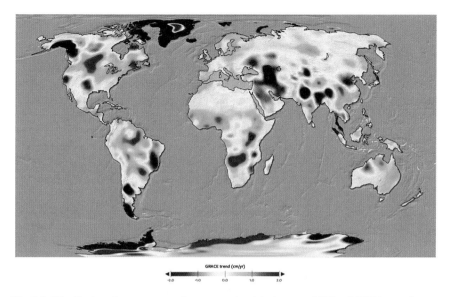

Fig. 1.2 Distribution of average mass change over the globe between 2004 and 2016 in cm of water equivalent per year (reddish tones: mass loss, bluish tones: mass gain). *Source* NASA (2020)

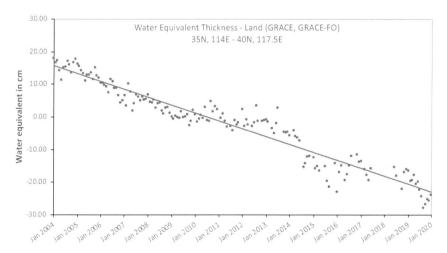

Fig. 1.3 Mass loss averaged over the North China Plain between 2004 and 2020 in monthly steps (in cm water equivalent) (Data retrieved from NASA's GRACE data analysis tool https://grace.jpl.nasa.gov/data-analysis-tool, note that no data were available mid 2017 to mid 2018 before the launch of GRACE FO)

On one hand the well yields decrease with declining groundwater levels, reaching zero, when the groundwater level falls below the level of the pump. On the other hand, the energy cost of pumping increases with depth to groundwater, making water more and more expensive. Eventually, pumping will have to decrease due to economic constraints and an equilibrium is reached again, however at much lower groundwater levels. In economics, optimal use of groundwater means maximizing net present value of revenue derived from it. This implies that depleting aquifer storage today yields a higher total benefit over time than preserving it for the future, as standard discounting values production gain by irrigation today higher than the same production gain in the future. Some economists (Gisser and Sanchez 1980a, b) have argued that compared to free market forces, contributions of groundwater management to social benefit are at best marginal.

This view undervalues the negative consequences of aquifer depletion encountered before a limit for pumping is reached. In particular, it undervalues water storage of an aquifer, which provides a buffer against droughts. Depleting an aquifer allows high agricultural production and leads to a grain bubble (Lester Brown in (George 2011)). A return to sustainable groundwater abstraction is almost inevitably associated with the bursting of this bubble and a possibly disruptive reduction in agricultural production.

An aquifer in dynamic equilibrium is characterized by its average recharge being equal to its average discharge in the long term. The equilibrium is disturbed, when the recharge decreases, or the discharge increases over a prolonged period of time. Consequently, groundwater levels will decline, eventually reaching a limit. There are basically three ways in which pumping can come to an end: The first is reaching the physical limit of drying up of the aquifer. The second is reaching the water quality limit. With increasing depth to groundwater, mineralization usually increases, and when it hits a threshold unsuitable for use, pumping will stop. The third, considered in the analysis of Gisser and Sanchez, is reaching the economic limit, in which case the water price due to cost of deep-well infrastructure and energy requirements becomes unacceptable. If groundwater levels develop towards any one of these limits, abstraction is not sustainable over time. The High Plains aquifer in Kansas is the first example of an aquifer, which has been exploited to the physical limit in certain parts (Whittemore et al. 2018).

Long before reaching any of the above limits, undesirable consequences of declining groundwater levels may arise, which require a reduction of pumping.

Adverse consequences may be related to **ecological concerns**. Groundwater cannot be viewed separately from surface water. When groundwater levels fall below a streambed, groundwater discharge to the stream stops and the groundwater-dominated dry-weather stream flow is depleted, possibly disrupted completely. Stream-depletion by groundwater table decline is a serious worldwide problem (Döll et al. 2009; de Graaf et al. 2014). Springs and wetlands suffer if their groundwater feed is cut off by consumptive uses (for water use terminology see Box 1.1). With groundwater table decline, phreatophytic plants relying on a shallow depth to groundwater may no longer be able to survive. The degradation of *populus euphratica* forests in the Tarim basin of Xinjiang is a vivid example for that phenomenon (Liu et al.

2005). Generally, shallow groundwater affects terrestrial ecosystems by sustaining river base-flow and root-zone soil water in the absence of rain. 22 to 32% of global land area is of this type (Fan et al. 2013). Its decline over time is a cause for concern.

Adverse consequences may also be related to **infrastructural safety and water quality**. A common phenomenon of declining groundwater tables is soil subsidence, which is prominent if soft aquifer layers are depressurized by pumping (Herrera-García et al. 2021). In Cangzhou (a city in Hebei province, NCP, China) for example, the soil subsidence has led to a sinking of the topography by 2.5 m. All over Hebei Province, soil fissures often several kilometers long have appeared (Gong et al. 2018). The most prominent example of soil subsidence due to groundwater pumping is Mexico City, where the settling has caused immense damage to infrastructure (Ortega-Guerrero et al. 1999). At the coast, groundwater level decline may lead to seawater intrusion as it is seen for example in California (Franklin et al. 2017), and the east coasts of India and China. In Hebei Province, the saline waterfront has advanced inland by up to 50 km due to groundwater level decline (CIGEM 2019) with the subsequent pollution of wells located within that zone.

Last but not least, adverse consequences may be related to **loss of resilience**. Storage volumes of aquifers are often large compared to surface water reservoirs and therefore able to buffer multi-year droughts. Depletion of an aquifer decreases its storage and thus its ability to buffer the stochastic nature of precipitation and surface water availability. Loss of storage makes a groundwater supply system—e.g. an irrigation system—less reliable. Maintaining storage by allowing recharge in years of good rains is considered an adaptation measure to extremes associated with climate change.

A disequilibrium between recharge and discharge can be caused by decreasing recharge (e.g. due to climate change), increasing discharge (due to increased pumping) or both at the same time. Typically, a climatic change with decreasing precipitation will both decrease recharge by rainfall and increase water demand for irrigation and thus discharge by pumping. Note that not every groundwater table decline indicates over-pumping. The aquifer could just be on the way from one equilibrium to another one, trading off discharge to streams or by phreatic evaporation against discharge by pumping. We generally speak of over-pumping, when in the long term—typically decades—discharge by pumping exceeds recharge.

Estimates of global unsustainable groundwater depletion vary from about 115 km^3/year (Döll et al. 2014) to 283 km^3/year (Wada et al. 2010) and 362 km^3/year (Pokhrel et al. 2012), with the respective authors' estimates of total global abstractions being anywhere between 600 and 1000 km^3/year. These figures are of interest to scientists who estimate for example the contribution of groundwater depletion to sea level rise or the impact of decreasing groundwater availability on the global production of agricultural goods. They are less relevant for water managers caring about a single aquifer, since groundwater is essentially a local resource, which must be managed locally.

Usage or consumption? Some water terminology

Volumes of water are expressed in m^3 or km^3, flowrates or pumping rates in m^3/s or m^3/yr.

If applied to the water delivered to our house or to an irrigated field, the colloquial word water consumption is inaccurate. Real "consumption" of water happens only when water is evaporated e.g. by the evapotranspiration of plants. This water is also not lost for the globe as it is recycled via the atmosphere. It may even be recycled locally in thunderstorms. However, usually it is lost for the catchment from which it is evaporated. In household and industry, the consumptive part of water use is small (<20%).

Water use in irrigation is mostly consumptive use (around 80%), of which again a part is productive use leading to plant growth and another part is non-productive evaporation from the soil. Irrigation water that is not consumed by evaporation or transpiration (around 20%) seeps to the underlying aquifer or flows to a drainage canal. Seepage can already happen during conveyance in an irrigation canal. Conveyance by piping reduces seepage.

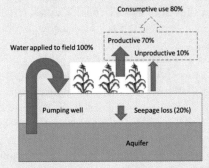

Depletion of an aquifer represents consumptive water use if seepage returns to the same aquifer. All water pumped from a deep aquifer, which does not receive irrigation backflow, counts as depletion. Irrigation with surface water provides a net recharge to the underlying aquifer by seepage. Water saving in agriculture only happens if consumptive use is reduced. A reduction in water demand due to a reduction in seepage losses in the field or during conveyance saves only little water (a fraction of the 20% seepage losses). It mainly saves energy by reducing the useless cycling of water by pumping

Water statistics of water authorities widely used in this report list total water use. Agronomists' figures usually express water consumption. A more comprehensive definition also counts water lost to sinks like the sea or rendered unusable by pollution as consumptive (Seckler 1996).

Box 1.1: Water terminology: usage versus consumption

The major user of groundwater by far is agriculture, which globally accounts for more than 70% of total water withdrawals and for more than 90% of total consumptive water use (Döll 2009). About 40% of irrigated agriculture relies on groundwater (Siebert et al. 2010). Its popularity is increasing, as it is a convenient resource, which is available throughout the year and at the location of use. Groundwater depletion due to domestic use also exists, but is usually confined to very large cities.

1.2 What Does Sustainable Groundwater Use Mean?

The most prominent sustainability problems in the field of water are related to groundwater. Besides the depletion issue, other concerns such as the reduction of low flows of rivers, the drying up of wetlands, seawater intrusion and soil salinization are related to groundwater levels (Alley et al. 1999; Kinzelbach et al. 2003; Liu et al. 2001).

Fig. 1.4   The red-lines
concept defining
the groundwater level range
required for sustainability

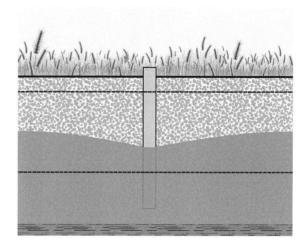

Sustainable groundwater use can be defined as an abstraction regime, which keeps groundwater levels within a range bounded by an upper and a lower limit—or two red lines—which guarantee the fulfilment of sustainability criteria specific for the region considered (Fig. 1.4).

The upper limit has the function of preventing phreatic evaporation in agricultural regions, which leads to water logging and salinization. The upper red line in that case is located at the extinction depth of phreatic evaporation, typically between 2 and 5 m below soil surface. The lower red line could be determined by ecological criteria such as low flow requirements of streams, or the maximum root depth of phreatophytic plants. In an agricultural context, it should incorporate the requirement of minimum well yield and a reserve required to overcome a design drought. Well yield determines the time needed to pump a given amount of irrigation water. If it drops below a critical minimum, it may limit the ability to provide sufficient water to crops when it is needed (Foster et al. 2017). Close to the coast, the lower red line is determined by the seawater level. Note that the red line levels are groundwater levels as observed in observation wells, possibly averaged over a certain time and area, and not the momentary dynamic water levels in a pumping well which may be considerably lower.

There are a number of best practices recommended by groundwater managers worldwide to control groundwater levels between the two red lines. Abstraction from an aquifer should allow the establishment of an equilibrium between recharge and discharge, which respects the two red lines, not at every moment, but on the average over times on the order of a decade. Surface water and groundwater should be used conjunctively. This means that surface water is the primary source of supply in years with average and above average rainfall and associated surface water flows, allowing groundwater to recharge, while in years of low flow or zero flow, groundwater takes over the supply. The ideal use of groundwater is as a buffer, with recharge in water rich years and drawdown of groundwater levels in dry years. Theoretically, this means that it is sufficient to design the maximum abstraction to stay below average recharge. However, in times of climate change, averages change, and management must be adaptive, correcting the average e.g. by considering moving averages or by actively controlling the groundwater level based on the red-line concept.

Real sustainability of agriculture irrigated with groundwater is achieved when water resources availability and agricultural production are in balance. Both an increase in supply and a decrease in demand can contribute to the restoration of an equilibrium between recharge and discharge in an over-pumped aquifer. This means that in the efforts to reach aquifer sustainability, an adaptation of the cropping system and a reduction of agricultural production should be taken as seriously as the search for new resources. In water scarce areas of today demand management of water resources increasingly replaces supply management.

An engineering measure of control is managed aquifer recharge (MAR), which uses excess water, e.g. surface water available outside of the irrigation season, for infiltration into the aquifer via ponds, canals, or wells. This water can then be pumped again at times of need. MAR is a promising adaptation measure to reduce vulnerability to climate change and hydrological variability. It can play a certain role in the restoration of the groundwater balance of aquifers. It can also be used to control saltwater intrusion or land subsidence. Finally, it can contribute to sustaining groundwater dependent ecosystems. The extreme form of MAR is water banking, where water is bought cheaply in times of excess, infiltrated into an aquifer, and pumped and sold in times of scarcity at the high water price associated with it. Several such schemes are working beneficially in the US and Australia. One of the first successful examples is the Arvin Edison groundwater bank in California (Scanlon et al. 2012).

Nevertheless, MAR is not a panacea. It is notoriously inefficient and plagued by clogging. It is always easier to pump water **out** of an aquifer than to put it back **into** the aquifer. Before considering such an operation the suitability of sites and methods, the costs of building and maintenance and the alternatives have to be investigated (Dillon et al. 2009). Given the order of magnitude of over-pumping In the North China Plain, the role of MAR in restoring the aquifer balance is limited. Two examples are discussed in Sect. 2.3.

1.3   Role of Irrigation in Over-Pumping in NCP

The NCP as defined in this book is bounded by the Taihang Mountains in the West, the Yan Mountains in the North, the Yellow River in the South and the Bohai Sea in the East (Fig. 1.5). Its area is about 140,000 km^2 and it is home to about 150 Mio. inhabitants. Due to its flat geomorphology, the NCP is one of the major agricultural areas in China. Within its boundaries, it produces 21% of China's wheat crop and 13% of its maize crop. Its climate is semi-arid with rainfall of about 500 mm/year and a potential evapotranspiration of about 1500 mm/year. There is a South-North gradient of precipitation, with higher rainfall in the South and lower rainfall in the North (Fig. 1.6). 70–80% of rainfall occurs in the summer monsoon season between

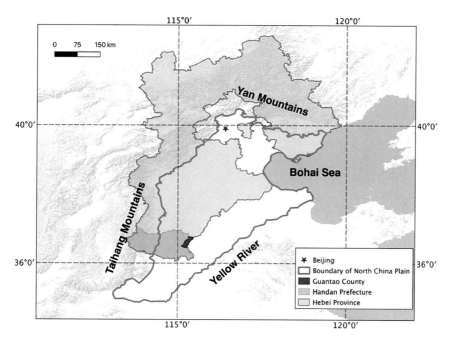

Fig. 1.5 The North China Plain (NCP) covering the sedimentary plain within the hydrogeological borders formed by the Yan mountains, the Taihang Mountains, the Yellow River and the Bohai Sea. It includes parts of Beijing, Tianjin, Hebei, Henan and Shandong Provinces

June and September.

The NCP can be divided into four hydrogeological zones: (I) the piedmont plain adjacent to the mountains in the west, (II) the central alluvial fan and lacustrine zone formed by the rivers coming from the Taihang and Yan mountains, (III) the flood plain created by the ancient Yellow River (including today's Zhang and Wei rivers) and (IV) the coastal plain bordering on the Bohai Sea in the East. Zones II and III occupy most of the NCP (Fig. 1.7).

The groundwater system of the NCP is divided into four main aquifers. The bottom of the first aquifer is at a depth of 40–60 m, the bottom of the second at 120–170 m, while the third and the fourth aquifers reach to depths of 250–350 m and 550–650 m, respectively. The first and the second aquifer are well connected and form the so-called "shallow aquifer", while the two deep layers are known as the "deep aquifer" (Wu et al. 2010). The shallow and the deep aquifer are only connected in the piedmont region, while with distance from the mountains they are increasingly separated by thick aquitards of low hydraulic conductivity and high salinity (Fig. 1.8).

The shallow aquifer has recharge from rainfall and to a lesser extent from irrigation water backflow and river infiltration. Natural groundwater recharge by rainfall is between 10 and 30% of precipitation (Fig. 1.6). The shallow aquifer's water quality in Zones II and III is usually not up to drinking water standard as mineralization is high (1–5 g/L TDS). The deep aquifer gets its recharge only in the piedmont region

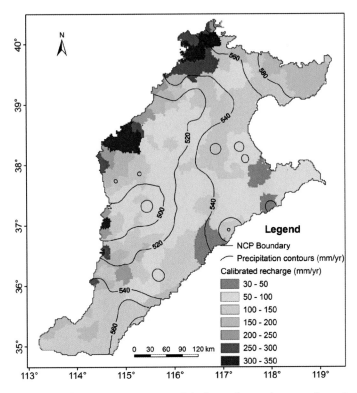

Fig. 1.6 Distribution of long-term average precipitation and groundwater recharge in the NCP. *Source* Cao et al. (2013)

(Zone I) on the western boundary. Its water satisfies drinking water standards with a mineralization usually well below 1 g/L. Large-scale irrigation in the NCP started in the 1950s and was based on surface water supplied through canals. The seepage losses led to a groundwater table rise and widespread soil salinization as the flat terrain did not provide an efficient drainage. This changed in the 1980s when China due to population growth had to supply more grain. The double cropping system of winter wheat and summer maize was promoted. The water demand of this crop rotation of about 900 mm/year surpasses the effective rainfall of about 500 mm/year. As rainfall is concentrated in summer, winter wheat has to rely almost completely on irrigation with groundwater (Fig. 1.9).

Thus, China's efforts to feed a growing population dramatically increased the use of groundwater since the 1980s, leading to widespread over-pumping. Typical declines in the shallow unconfined and the deep confined aquifers are shown in Figs. 1.10 and 1.11. The decline of groundwater levels in the shallow aquifer (Fig. 1.10) was fast in the 1990s and slowed down after 2000. In recent years, levels are relatively constant, apart from the seasonal variation indicating the abstraction of irrigation water. In some regions, even a rise has been observed. The same tendency

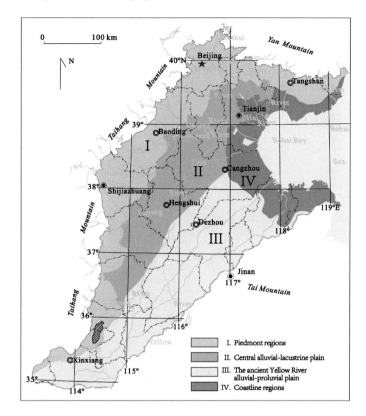

Fig. 1.7 Geological map of the NCP. *Source* Meng et al. (2015)

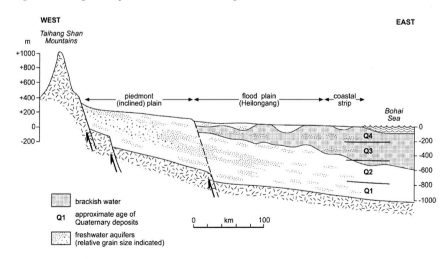

Fig. 1.8 Hydrogeological cross section of the NCP. *Source* Foster et al. (2004)

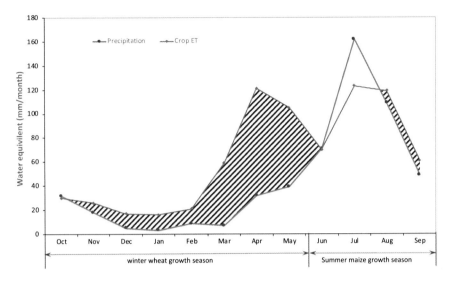

Fig. 1.9 Distribution of rainfall, crop water demand (crop ET) and water deficit (hatched area) over the year in the double cropping system of winter wheat and summer maize (Average data for Guantao County)

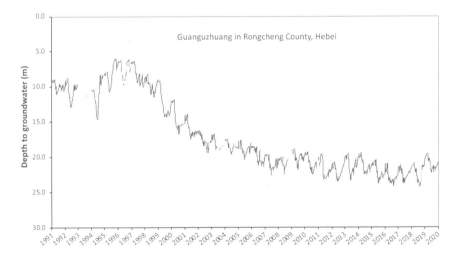

Fig. 1.10 Groundwater table decline in the shallow aquifer, using observation well in Guan-guzhuang, Rongcheng County, Hebei as an example

is seen in an average of 559 unconfined aquifer wells distributed over the NCP (Zhang et al. 2020). The authors ascribe local rises after 2014 to the influence of water imports by the South North Water Transfer scheme (SNWT). The decline in the 1980s and 1990s is not caused by pumping alone. It is also influenced by a

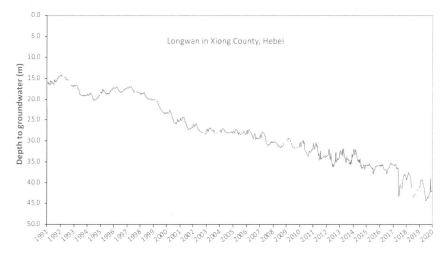

Fig. 1.11 Decline of piezometeric head in deep aquifer, using the example of an observation well located in Fuwangcun, Xiong County, Hebei

decrease in rainfall (Fig. 1.12). After 2003, the decline rate of shallow groundwater levels decreased. Zhang et al. (2017) estimate that the increase of rainfall after 2002, contributed with 64% to that slowdown, while water saving contributed with 36%. Ignoring the influence of precipitation in the interpretation of groundwater levels may lead to an overestimation of the efficacy of water saving efforts.

The piezometric levels of the deep aquifer (Fig. 1.11) are not related to local precipitation as the aquifer receives little local recharge. The levels have been falling

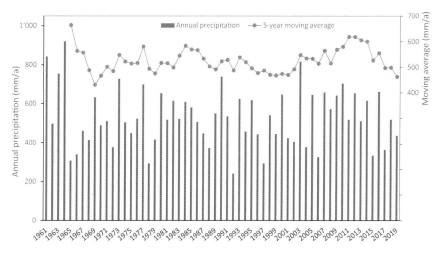

Fig. 1.12 Development of rainfall since 1961 (Example Guantao County). Tendencies are clarified by taking a 5-year moving average

continuously since the nineties due to pumping of water not only for irrigation but also for households and industry. The decline even accelerated in recent years, which is also seen by an increasing seasonal amplitude.

The head difference between the two aquifers is ever increasing, which conjures up the danger of polluting the deep aquifer by saline water intruding from the aquiclude. This danger has previously been indicated by Foster et al. (2004) and may prevail much longer than the present over-pumping issue. The processes involved are depicted in Fig. 1.13. In the NCP, evidence of increasing leakage recharge from the shallow aquifers to the deep aquifers has been found after 2000 (Cao et al. 2013; Huang et al. 2015). While the quantities are far too small to effectively recharge the deep aquifer, they present a danger for its water quality.

The distribution maps of groundwater levels and piezometric heads (Figs. 1.14 and 1.15) show the overexploitation zones as pronounced cones of depression, both

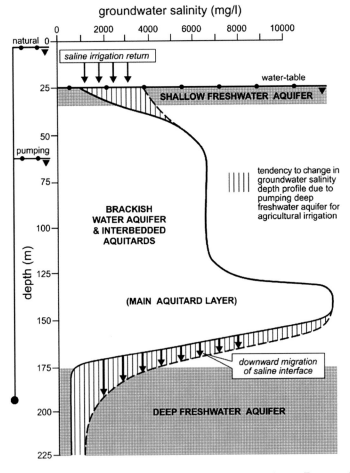

Fig. 1.13 Processes leading to salinization of groundwater resources. *Source* Foster et al. (2004)

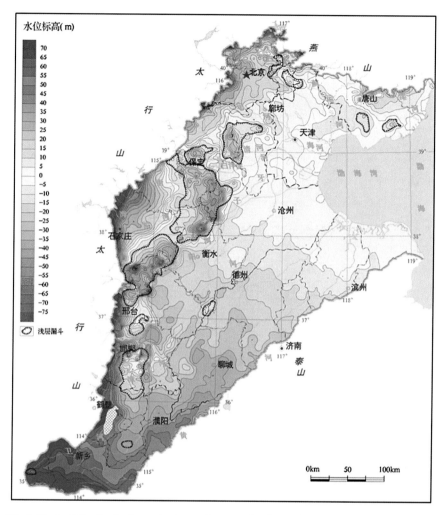

Fig. 1.14 Spatial distribution of observed water levels in the shallow aquifer in the North China Plain in 2019 (Cones of depression marked with red contours) (Groundwater levels in m asl, in steps of 5 m from –75 m asl to 70 m asl). *Source* CIGEM (2019)

in the shallow and the deep aquifers. The over-pumping is clearly more serious in the deep aquifer, even though its abstractions are only about a quarter of the abstractions in the shallow aquifer. It is a confined aquifer, which consequently has a very small storage coefficient. Therefore, a relatively small abstraction already causes a large drawdown. The deep aquifer is not only used by agriculture but also by households and industry because of its drinking water quality.

A rough water balance has been suggested in various groundwater-modelling efforts. Results for the period 2002 to 2008 from (Cao et al. 2013) and for 2002 to 2003 from (Wang et al. 2008) are shown in Table 1.1. Note that fluxes are not unique,

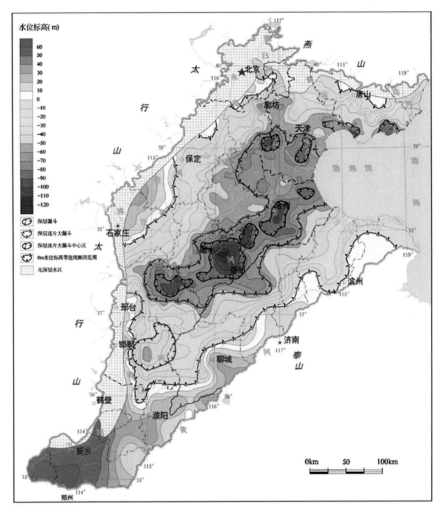

Fig. 1.15 Spatial distribution of observed piezometric levels in the deep aquifer in the North China Plain in 2019 (Cones of depression have joined to a large depression zone. Drawdown zones of different intensity are marked with red and black contour lines) (Piezometric levels in m asl, in steps of 10 m from –120 m asl to 60 m asl). *Source* CIGEM (2019)

even if they manage to reproduce observed head changes.

The depletion of storage in both models is close to the estimate made based on GRACE data (Fig. 1.3). It indicates the gap between discharge and recharge, which has to be closed in order to return to a sustainable mode of resource use. Even if abstractions could be reduced to achieve zero storage change within a few years, the restoration of groundwater levels requires much longer times. A storage change of zero would freeze present average groundwater levels. Only an excess of recharge over discharge, which is bound to be small, would allow to gradually refill

Table 1.1 Water balance of the NCP aquifer system

Item	Flux (in km^3/year) (+in, −out) average 2002–2008 (Cao et al. 2013)	Flux (in km^3/year) (+in, −out) average 2002–2003 (Wang et al. 2008)
Recharge by precipitation and irrigation backflow	+18.71	+20.77
Boundary flux and river seepage	+2.21	+2.60
Pumpage (shallow aquifer)	−17.70	−17.52
(deep aquifer)	−5.01	− 3.11
Phreatic evaporation and outflow to the sea	−2.58	−6.32
Storage change	−4.37	−3.58

the accumulated depletion of 40 years. A return to the shallow aquifer's groundwater levels of 1980 is not desirable as they were so high that they led to water logging and soil salinization.

While the global consideration of Table 1.1 gives a feeling for the order of magnitude of the problem and the efforts needed in its solution, it is not useful for the planning of concrete management measures. Figures 1.14 and 1.15 show clearly that aquifer depletion is not distributed homogeneously over the NCP. They also show that the situation of the deep aquifer is more severe than the situation of the shallow aquifer. The adverse consequences of long-term groundwater overexploitation such as land subsidence, ground fissures, rivers drying up, wetland degradation and seawater intrusion are also area specific. Finally, zero storage change for the whole aquifer system could hide deterioration in some locations by improvement in others. This means that for the efficient restoration of a sustainable pumping regime a thorough, spatially resolved analysis is necessary. For an example of such a local analysis, we turn to the pilot area of Guantao County, which is a typical county in the NCP.

1.4 Requirements for Sustainability in NCP and Guantao as an Example

Guantao County, located in Handan Municipality of Hebei Province, is a typical county of the NCP (Figs. 1.5 and 1.16). It was chosen as pilot region for the implementation of the Sino-Swiss groundwater project. It had been the object of an earlier study funded by the World Bank under a GEF project (Foster and Garduño 2004). Its long-term average annual precipitation is 525 mm with an annual potential evapotranspiration of 1516 mm. Annual average temperature is 13.4 °C. Guantao has a population of 363,000, an area of 456 km^2 and an irrigated agricultural area of

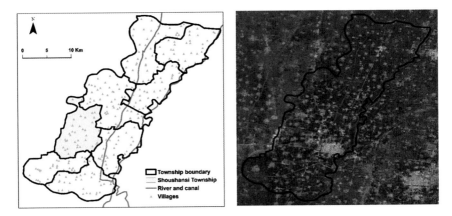

Fig. 1.16 Guantao County map (left) and remote sensing image (right)

almost 300 km^2. The total annual water use is about 123 Mio. m^3, of which agricultural irrigation covers 82%, industrial water use 4%, domestic and other water use 14%.

The main planting area is dedicated to the double cropping of winter wheat and summer maize. The main crops, their planting areas, and calendars as well as their irrigation norms are listed in Table 1.2.

The crop water supply depends on three sources: precipitation, the exploitation of shallow and deep groundwater layers through about 7300 shallow and 300 deep wells, and surface water provided both from the Weiyun River and from the Yellow River via Weidaguan Canal and Minyou Canal. For household and industry, the SNWT scheme has been providing 4 Mio. m^3 per year since 2014. This allowed a reduction of deep-aquifer abstractions for households and industry by about one half.

Wells are operated and managed by well managers, who are owners of a well or take care of a well co-owned by several households. A household of four persons is allocated a planting area of about 1/3 of a hectare. A well typically covers about 3 ha of cropland. All wells are powered by electricity and equipped with an electricity meter. Village electricians collect electricity fees based on the readings of electricity meters by well managers, who organize the recording of irrigation electricity use and collect electricity fees from farming families sharing the same well. The collected fees include the rural, subsidized electricity fee (0.5115 CNY/kWh) for pumping and in some regions an additional fee (about 0.3–0.5 CNY/kWh depending on the farmland's distance from the well) charged for the service of the well manager.

Table 1.2 Main crops and their respective planting area, growth period, planting month, harvesting month and irrigation norm. (Data supplied by Guantao Department of Water Resources based on the local year book for 2012.)

Crop type	Total planting area (ha)	Growth period (days)	Planting month	Harvest month	Irrigation norm (mm)
Winter wheat	20,570	240	October	June	247.5
Summer maize	16,663	120	June	October	75
Cotton	5,730	210	April	November	150
Oil crop (soy)	1,025	120	June	October	150
Oil crop (peanut)	3,761	160	April	October	90
Tomato	325	60	March	May	300
Tree garden	1,000				
Orchard (apple)	1,615	210	March	October	180
Orchard (pear)	500	210	March	October	225
Orchard (wine)	350	100	April	September	150
Orchard (peach)	200	100	March	July	180
Millet	157	120	June	October	75
Vigna radiate	33	100	April	August	75
Fall sweet potato	211	120	June	October	150
Vegetables	7,675	140	February	July	375
Melons	183	120	April	July	375

Note Due to double cropping, the sum of planting areas is larger than Guantao's agricultural area

To reach the goal of sustainable groundwater use, a management system has been set up in Guantao, integrating policy with data-based decision support. It comprises three modules: data monitoring, decision support by modelling, and implementation of policies in the field (Fig. 1.17). Monitored items include groundwater levels, pumping rates, and land use. In addition, data on surface water imports and meteorological quantities are collected. All data are transferred to a server, where they are analyzed by various models and policy options are identified. The chosen policies are then implemented in the field. Success or failure can be assessed via the monitoring data in the following year. The management cycle is gone through annually, decisions being taken before the winter wheat planting in October.

The following chapters demonstrate how such a management system can be successfully implemented. This includes a review of the policy options used in

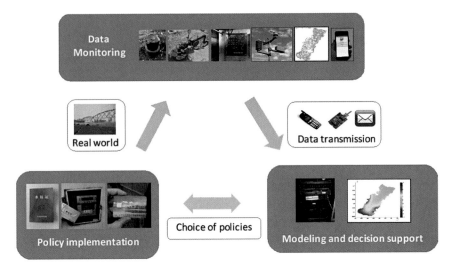

Fig. 1.17 Architecture of a real-time groundwater management system, including real-time data monitoring and transmission, modelling for decision support, and policy implementation in the field, with feedback by data monitored

the NCP to control groundwater over-pumping, the evaluation of the most effective control measures in the field, the farmers' feedback obtained from a household survey and a policy test based on game playing. Eventually the above-mentioned management system was set up in Guantao through an integrated decision support platform.

References

Alley WM, Reilly TE, Franke OL (1999) Sustainability of groundwater resources. United States Geological Survey Circular 1186

Cao G, Zheng C, Scanlon BR, Liu J, Li W (2013) Use of flow modeling to assess sustainability of groundwater resources in the North China Plain. Water Resour Res 49(1):159–175. https://doi.org/10.1029/2012WR011899

CIGEM (2019) Investigation and Assessment of Water Resources in China. Report of the China Institute of Geo-Environmental Monitoring, Beijing

de Graaf IEM, van Beek LPH, Wada Y, Bierkens MFP (2014) Dynamic attribution of global water demand to surface water and groundwater resources: effects of abstractions and return flows on river discharges. Adv Water Resour 64:21–33. https://doi.org/10.1016/j.advwatres.2013.12.002

Dillon P, Pavelic P, Page D, Beringen H, Ward J (2009) Managed aquifer recharge. Waterlines Report Series No. 13, National Water Commission of Australia. ISBN: 978-1-921107-71-9

Döll P (2009) Vulnerability to the impact of climate change on renewable groundwater resources: a global-scale assessment. Environ Res Lett 4:035006. https://doi.org/10.1088/1748-9326/4/3/035006

Döll P, Fiedler K, Zhang J (2009) Global-scale analysis of river flow alterations due to water withdrawals and reservoirs. Hydrol Earth Syst Sci 13:2413–2432. https://doi.org/10.5194/hess-13-2413-2009

Döll P, Müller-Schmied H, Schuh C, Portmann FT, Eicker A (2014) Global-scale assessment of groundwater depletion and related groundwater abstractions: Combining hydrological modeling with information from well observations and GRACE satellites. Water Resour Res 50:5698–5720. https://doi.org/10.1002/2014WR015595

Fan Y, Li H, Miguez-Macho G (2013) Global patterns of groundwater table depth. Science 339(6122):940–943. https://doi.org/10.1126/science.1229881

Feng W, Zhong M, Lemoine JM, Biancale R, Hsu HT, Xia J (2013) Evaluation of groundwater depletion in North China using the Gravity Recovery and Climate Experiment (GRACE) data and ground-based measurements. Water Resour Res 49:2110–2118. https://doi.org/10.1002/wrcr.20192

Foster S, Chilton P (2003) Groundwater: the processes and global significance of aquifer degradation. Philos Trans R Soc Lond Ser B 358:1957–1972. https://doi.org/10.1098/rstb.2003.1380

Foster S, Garduño H (2004) China: towards sustainable groundwater resource use for irrigated agriculture on the North China Plain. World Bank. Sustainable groundwater management: lessons from practice. GW-MATE Case Profile Collection Number 8

Foster S, Garduño H, Evans R, Olson D, Tian Y, Zhang W, Han Z (2004) Quaternary Aquifer of the North China Plain—assessing and achieving groundwater resource sustainability. Hydrogeol J 12:81–93. https://doi.org/10.1007/s10040-003-0300-6

Foster T, Brozovic N, Butler AP (2017) Effects of initial aquifer conditions on economic benefits from groundwater conservation. Water Resour Res 53:744–762. https://doi.org/10.1002/2016WR019365

Franklin H et al (2017) Recommendations to address the expansion of seawater intrusion in the Salinas Valley groundwater basin. Monterey County Water Resources Agency, Special Reports Series 17-01

George A (2011) Lester Brown: The food bubble is about to burst. New Scientist 209(2798):27. https://doi.org/10.1016/S0262-4079(11)60259-5

Gisser M, Sánchez DA (1980) Competition versus optimal control in groundwater pumping. Water Resour Res 16:638–642. https://doi.org/10.1029/WR016i004p00638

Gisser M, Sánchez DA (1980) Some additional economic aspects of ground water resources replacement flows in semi-arid agricultural areas. Int J Control 31(2):331–334. https://doi.org/10.1080/00207178008961044

Gong H, Pan Y, Zheng L, Li X, Zhu L, Zhang C, Huang Z, Li Z, Wang H, Zhou C (2018) Long-term groundwater storage changes and land subsidence development in the North China Plain (1971–2015). Hydrogeol J 26:1417–1427. https://doi.org/10.1007/s10040-018-1768-4

Herrera-García G, Ezquerro P, Tomás R, Béjar-Pizarro M, López-Vinielles J, Rossi M, Mateos RM, Carreón-Freyre D, Lambert J, Teatini P, Cabral-Cano E, Erkens G, Galloway D, Hung W, Kakar N, Sneed M, Tosi L, Wang H, Ye S (2021) Mapping the global threat of land subsidence. Science 371(6524):34–36. https://doi.org/10.1126/science.abb8549

Huang Z, Pan Y, Gong H, Yeh PJF, Li X, Zhou D, Zhao W (2015) Subregional-scale groundwater depletion detected by GRACE for both shallow and deep aquifers in North China Plain. Geophys Res Lett 42:1791–1799. https://doi.org/10.1002/2014GL062498

Kinzelbach W, Brunner P, Bauer-Gottwein P, Siegfried T (2003) Sustainable groundwater management. Episodes 26(4):279–284

Konikow LF, Kendy E (2005) Groundwater depletion: a global problem. Hydrogeol J 13:317–320. https://doi.org/10.1007/s10040-004-0411-8

Liu C, Yu J, Kendy E (2001) Groundwater exploitation and its impact on the environment in the North China Plain. Water Int 26(2):265–272. https://doi.org/10.1080/02508060108686913

Liu J, Chen Y, Chen Y, Zhang N, Li W (2005) Degradation of populous Euphratica community in the lower reaches of the Tarim River, Xinjiang China. J Environ Sci (China) 17(5):740–747

MEA (2005) Millennium Ecosystem Assessment, 2005. Ecosystems and Human Well-being: Synthesis. Island Press, Washington, DC

Meng S, Liu J, Zhang Z, Lei T, Qian Y, Li Y, Fei Y (2015) Spatiotemporal evolution characteristics study on the precipitation infiltration recharge over the past 50 years in the North China Plain. J Earth Sci 26:416–424. https://doi.org/10.1007/s12583-014-0494-7

NASA (2020) https://grace.jpl.nasa.gov/data/get-data/

Ortega-Guerrero A, Rudolph DL, Cherry JA (1999) Analysis of long-term land subsidence near Mexico City: field investigations and predictive modeling. Water Resour Res 35(11):3327–3341. https://doi.org/10.1029/1999WR900148

Pokhrel YN, Hanasaki N, Yeh PJF, Yamada TJ, Kanae S, Oki T (2012) Model estimates of sea-level change due to anthropogenic impacts on terrestrial water storage. Nat Geosci 5:389–392. https://doi.org/10.1038/ngeo1476

Rodell M, Famiglietti JS, Wiese DN et al (2018) Emerging trends in global freshwater availability. Nature 557:651–659. https://doi.org/10.1038/s41586-018-0123-1

Scanlon BR, Faun CC, Longuevergne L, Reedy RC, Alley WM, McGuire VL, McMahon PB (2012) Groundwater depletion and sustainability of irrigation in the US High Plains and Central Valley. PNAS 109(24):9320–9325. https://doi.org/10.1073/pnas.1200311109

Siebert S, Burke J, Faures JM, Frenken K, Hoogeveen J, Döll P, Portmann FT (2010) Groundwater use for irrigation—a global inventory. Hydrol Earth Syst Sci 14:1863–1880. https://doi.org/10.5194/hess-14-1863-2010

Seckler DW (1996) The new era of water resources management: from "dry" to "wet" water savings. IWMI Report

Wada Y, van Beek LPH, van Kempen CM, Reckman JWTM, Vasak S, Bierkens MFP (2010) Global depletion of groundwater resources. Geophys Res Lett 37:L20402. https://doi.org/10.1029/2010GL044571

Wang S, Shao J, Song X, Zhang Y, Huo Z, Zhou X (2008). Application of MODFLOW and geographic information system to groundwater flow simulation in North China Plain. China Environ Geol 55(7):1449–1462. https://doi.org/10.1007/s00254-007-1095-x

Whittemore DO, Butler JJ Jr, Brownie Wilson B (2018) Status of the High Plains Aquifer in Kansas. Kansas Geological Survey Technical Series 22

Wines M (2013). Wells dry, fertile plains turn to dust. New York Times, May 19, 2013.

Wu A, Li C, Xu Y et al (2010) Key issues influencing sustainable groundwater utilization and its countermeasures in North China Plain. South North Water TransfS Water Sci Technol 8(6):110–113 (In Chinese)

Zhang HB, Singh VP, Sun DY, Yu QJ, Wei C (2017) Has water-saving irrigation recovered groundwater in the Hebei Province plains of China? Int J Water Resour Dev 33(4):534–552. https://doi.org/10.1080/07900627.2016.1192994

Zhang C, Duan Q, Yeh PJF, Pan Y, Gong H, Gong W, et al (2020) The effectiveness of the South-to-North Water Diversion Middle Route Project on water delivery and groundwater recovery in North China Plain. Water Resour Res 56:e2019WR026759. https://doi.org/10.1029/2019WR026759

Chapter 2
Policy Options of Over-Pumping Control in the NCP

Starting in the 1990s, China has been issuing regulations and policy rules related to groundwater management and pumping control on both national and provincial levels. These policies include the requirement of permits for well drilling, a well spacing policy, pumping quota management, water resources fee collection, setting of irrigation water prices, a water rights system, water markets, and more. Since the early 2010s, the central government increasingly paid attention to the groundwater depletion issue, leading to the deployment of the pilot program "Comprehensive Control of Groundwater Overdraft in North China Plain" in 2014. It coordinated efforts of several ministries and included innovative measures such as subsidies for fallowing of winter wheat and substitution of groundwater by surface water, especially for household and industry through the South-North Water Transfer Project. In Guantao County about 8 Mio. m3/yr of groundwater were saved through subsidized fallowing of winter wheat, while the import of surface water could be increased to almost 60 Mio m3/yr. However, continued funding of subsidies for fallowing is not guaranteed and the seemingly large imports of Yellow River water were used very inefficiently as the region lacks storage facilities to cope with water arriving off-season.

In China, policies are generally implemented through a top-down approach. Even though some policy pilots start on provincial or county level, eventually a national policy follows, upscaling the pilot experience to all relevant regions. This chapter summarizes China's recent national policies on groundwater management and assesses the efficiency of major measures for groundwater over-pumping control implemented through various pilots in Hebei province from 2014 to 2017. The chapter also highlights the most effective control measures adopted for further implementation in Beijing-Tianjin-Hebei (BTH) region from 2018 on and their implementation status. The governance structure of the water sector in China is introduced presenting the overall picture of groundwater management across different governmental bodies.

© The Author(s) 2022   25
W. Kinzelbach et al., *Groundwater Overexploitation in the North China Plain:*
A path to Sustainability, Springer Water,
https://doi.org/10.1007/978-981-16-5843-3_2

2.1 China's Groundwater Policies in Recent years

China has issued regulations and policy rules related to groundwater management and pumping control on both national and provincial levels. These policies include the requirement of permits for well drilling, a well spacing policy, pumping quota management, water resources fee collection, setting of irrigation water prices, a water rights system, water markets, and more. A summary of these policy instruments is given below:

2.1.1 Permit Policy for Well Drilling

In the early 1990s, some provinces in northern China began to implement formal or informal well-drilling permit policies to control the drilling of tube wells and thus limit the utilization of groundwater. Although the policy has been implemented in certain provinces as early as the 1990s and is effective up to now, well permits were not included in the 2002 Water Law and no national water regulation has addressed this policy. Shen (2015) pointed out that well permits are inconsistent with the recent reform aiming at a reduction of administrative permits in China.

Based on a survey covering six provinces (Hebei, Henan, Shanxi, Shaanxi, Liaoning and Inner Mongolia) in North China, done by Prof. Jinxia Wang and her team at Peking University, in 1995 18% of villages had implemented a well-drilling permit policy. This share increased to 34% by 2004 and to 54% by 2015 (Wang et al. 2020a).

2.1.2 Well-Spacing Policy

In 2010, the Ministry of Water Resources issued technical guidance on tube well spacing in both rural and urban regions (MWR 2010). Depending on the distance, the groundwater withdrawal of one tube well will affect the groundwater availability at other tube wells in the same aquifer (Huang et al. 2013). Frija et al. (2015) stated that management tools such as appropriate well spacing are needed in areas with groundwater overexploitation and degradation. Therefore, a well-spacing policy to control farmers' tube well drilling is a crucial measure to ensure the supply reliability of groundwater irrigation. However, it is both time-consuming and technically demanding to provide scientific information for well spacing based on the local hydrogeological conditions. The implementation of this policy so far depends mainly on local people's experience (e.g. county officials, drilling teams, farmers).

2.1.3 Quota Management

Water quota management was first introduced in China in the 2002 "Water Law". It only became a priority policy instrument after the "Three Red Lines" policies were issued by the central government in 2012 (State Council 2012). The central government required river basin management authorities and local water resources bureaus to determine water quotas for various water users at different administrative levels (i.e. river basin, province, city, county, irrigation district, and village). Under the water quota system, all water users should be issued withdrawal permits from upper-level water management authorities and their withdrawal rates (of both surface and groundwater) should not exceed the allocated quota. In recent years, the water quota concept has also been used in groundwater pumping control in the framework of the agricultural water pricing reform.

Despite central and local governments' efforts, implementation of water quota management remains difficult in rural areas. The main reasons are: First, water use differs by crops, farmers, and regions (according to soil and climate conditions) and it is difficult to calculate generally applicable water quota. Second, metering facilities for groundwater use in irrigation rarely exist. Third, the water rights system has not yet been established in all of China, and the relationship between water quota and this system remains unclear.

2.1.4 Water Resources Fee and Tax

(Kemper 2007) indicated that a water resources fee could provide incentives to use groundwater more efficiently. This is especially true if this fee is tied to the volume of groundwater used. Since the early 1980s, water resources fees have been introduced in certain provinces in northern China (e.g. Tianjin, Shanxi, Beijing) (Shen 2015). They are also included in the 2002 Water Law. In 2006 China's central government issued the "Regulations on the Management of Water Abstraction Permit and Collection of Water Resources Fee". The fee is collected by the water administration departments at the county level based on approved water abstraction permits. Where an abstraction permit is approved by river basin management organizations, the fee is collected by the relevant department of the province in which the water intake is located. It is based on the actual water abstraction volume and fee standards (Shen 2015).

Collecting groundwater resource fees in rural areas remains a major challenge for policymakers. Groundwater cost for irrigation includes pumping costs (including cost for drilling and equipment of tube wells, electricity cost or diesel fuel cost) but does not account for the scarcity value of groundwater. As (Yu et al. 2015) pointed out, the water price does not include the groundwater resource fee in most parts of rural China.

There is an ongoing transformation from a water resources fee to a water resources tax to enhance its potential role. In 2016 Hebei Province was selected as the pilot province for the fee to tax transformation. From 2017 the fee to tax reform has been expanded to 9 provinces including Beijing, Tianjin, Shanxi, Inner Mongolia, Henan, Shandong, Sichuan, Ningxia, and Shaanxi. No matter whether fee or tax, it is difficult to implement the policy in rural groundwater use. Farmers' low revenue from agricultural production and the lack of metering facilities for millions of irrigation pumping wells are both huge obstacles preventing the collection of a water resources tax from individual farmers.

2.1.5 Irrigation Water Price Policy

Over the past 40 years, the reform of China's irrigation water price policy has made some progress. After the first water fee regulation in 1985, surface water supply for irrigation was transformed from being fully subsidized to covering at least part of the cost by a supply fee. In 1992, price bureaus took over the management responsibilities for the irrigation fee from the Water Resources Bureaus, changing the nature of the fee from an administrative issue to a commodity issue. The irrigation fee further changed from a single to a two-part structure in the last two decades. The two parts include a basic fee charged by area and a volumetric fee charged by the amount of water used. In the last decade, the scarcity value of water resources was added as a component of the irrigation fee. In addition, from 2016 on, the central government began selecting regions to set up pilot projects to shift from water resources fees to water resources taxes. This might seem a change in name only. However, fees go to the water administration, possibly reducing their interest in water saving, while taxes go to the government tax office. The government also realized that irrigation price reform should be accompanied by improvement of irrigation facilities and institutional innovation (such as establishing WUAs).

Reform progress was mainly related to the price of surface water resources for irrigation, and not to the price of groundwater. In groundwater irrigation the major investors are farmers or village collectives. They mainly pay electricity or diesel fuel cost incurred for pumping water and so far, do not need to pay the resource fee. Collection of a groundwater resource fee is what the government expects from the reform of the irrigation water price. However, due to high implementation cost for collecting such a fee, until now this expectation cannot be fulfilled. Therefore, groundwater irrigation cost presently mainly consists of the energy cost (electricity or diesel fuel cost), cost for drilling and equipment of tube wells and eventually labor or service cost incurred during irrigation.

Due to poor measurement facilities and high implementation cost, it is hard to implement volumetric groundwater irrigation fees in the field. However, since most tube wells include electricity metering, and the major operation cost of a tube well is the electricity cost, it is common to collect groundwater irrigation fees based on electricity use. Charging groundwater irrigation fees according to electricity use can be treated as a proxy approach to a volumetric irrigation fee. A field survey by (Wang et al. 2020b) tracked 125 community tube wells managed by well managers.

For this sample, the percentage of tube wells collecting a groundwater irrigation fee proportional to the electricity fee increased from 36% in 2001 to 78% in 2015. In another approach, tube well managers collected groundwater irrigation fees by time of use of the well, which is another type of proxy approach to a volumetric irrigation fee. Over the past 15 years, the share of tube wells collecting groundwater irrigation fees by time of use slightly decreased from 25% in 2001 to 22% in 2015. Before 2015, some tube well managers also collected groundwater irrigation fees by area irrigated or by the amount of diesel fuel consumed.

2.1.6 Water Rights System and Water Markets

The Chinese government has been trying to set up a water rights system and allocate water through market mechanisms since the early 2000s (Calow et al. 2009). In theory, regulatory caps on total water use within a given region in a rational water rights system can lead to socially optimal allocation of water resources; thus, water can be used by those who value it most (Howe et al. 1986), (Debaere et al. 2014). Considering the potential benefit, the central government has been issuing regulations to promote the development of a water rights system over the last two decades. The first two important regulations were issued in 2005: "Some Opinions on Water Rights Transfer and Establishing a Framework of Water Rights System". In 2014, the government launched formal pilot projects in seven provinces to further accelerate the development. These provinces included Ningxia, Jiangxi, Hubei, Inner Mongolia, Henan, Gansu, and Guangdong. To support the implementation of pilot projects and encourage water rights transactions among regions, sectors, and individual water users, the MWR issued the "Temporary Management Regulation on Water Rights' Transfer" in 2016. In the same year, the first national Transaction Institute of Water Rights was established in Beijing.

So far, only a few pilot projects of water rights transfer have been considered successful, especially those which trade among regions and industries. For example, there are two prominent water rights transfer projects (Speed 2009), (Moore 2015): between Dongyang and Yiwu in Zhejiang province, and between agricultural and industrial sectors in Inner Mongolia and Ningxia provinces. Though successful, they are mainly coordinated by local governments; water users in these regions have not been directly involved in the transactions. It should also be noted that while the transfer of water rights may increase water use efficiency it does not reduce the amount of water pumped from an aquifer and is therefore no solution to over-pumping.

It is more difficult to establish a water rights system to promote rights transfer among irrigation water users in rural areas. At the irrigation district (ID) level, water rights have only been granted to farmers at a few select pilot sites, where water rights transactions among individual farmers are not always effective (Sun et al. 2016). Field surveys in China seldom found evidence of water rights' transfer among farmers. In fact, many farmers are unaware of their water use rights or the fact that they can be traded. A typical example for establishing a water rights system in rural areas is the institutional reform in Zhangye Prefecture in Gansu Province. Here, water rights have been granted to individual farmers in the form of water rights certificates. These certificates state the upper limit of the amount of water a household can buy, which is computed by area and crop irrigation quota. Even so, transactions involving water rights are rare in Zhangye. Importantly, because of poor implementation and high monitoring cost, water rights certificates do not have a sustainable function in reducing irrigation water demand. They only played a significant role in the early stages of the reform, where irrigation of wheat was reduced by 23% (before 2010). The survey (Sun et al. 2016) also found that farmers incurred practically no penalty for exceeding their water rights, which encourages them to use yet more water.

Despite progress made in establishing a water rights system and developing water markets, China still faces challenges in expanding reforms. There has also been a heated debate on the suitability of water markets in rural areas. The major issue is that initial water rights have not yet been allocated to various water users in most regions (Wang et al. 2017). It is impossible to develop water markets without a fully established water rights system. Recently, a water quota system has been suggested for allocating initial water rights to users. However, there is no clear agreement on the relationship between the water rights system and the water quota policy. In addition, the implementation of a water quota policy in rural areas has been slow because of the lack of metering facilities and the high cost involved in monitoring large numbers of small-holder farmers' wells. Therefore, some officials and scholars question its suitability for developing water markets in rural China, at least at the individual farmer level. If possible, it is better to encourage trade at the level of WUAs or IDs instead. (Lewis and Zheng 2018) noted that promoting water trade at the WUA level requires strong efforts to encourage farmers to participate in the activities of WUAs. Finally, the potential effects of water rights transfer on disadvantaged water users and on the environment also need to be seriously considered (Heaney et al. 2005; Johansson et al. 2002; Etchells et al. 2006).

2.1.7 National Policy Focus: NCP's Groundwater Over-Pumping

Regardless of all the policy rules and regulations published, implementation lags and groundwater over-pumping remains a major issue in the NCP. It has become a focus of national policy, appearing in many important state documents issued by the Central Committee of the Communist Party of China and the State Council. The most recent state documents include:

1. **"Decision of the Central Committee of the Communist Party of China and the State Council on Accelerating Water Conservancy Reform and Development" in 2011**

   This document demands that groundwater over-pumping should basically be stopped by 2020.

2. **"Opinions of the State Council on Implementing the Strictest Water Resources Management System" in 2012**

   This document calls for strict groundwater management and protection to reach a balance between groundwater exploitation and recharge.

3. **"Decisions of the Central Committee of the Communist Party of China on Several Major Issues of Comprehensively Deepening Reform" in 2013**

   This document proposes several measures for groundwater management, including the establishment of pre-warning mechanisms based on the monitoring of resources and environmental carrying capacity and the implementation of restrictive measures for areas where water and soil resources are exploited beyond their capacities. The measures include adjustment of cropland area in regions with severe pollution and severe groundwater overexploitation, in order to rehabilitate arable land, rivers and lakes.

4. **"Opinions of the Central Committee of the Communist Party of China and the State Council on Accelerating the Construction of Ecological Civilization" in 2015**

   The document states that the balance between groundwater extraction and recharge should gradually be reached by implementing groundwater protection and comprehensive management of the over-exploited areas characterized by groundwater depression cones.

5. **"Outline of the 13th Five-Year Plan for the National Economic and Social Development of the People's Republic of China" in 2016**

   The document states that groundwater management should focus on areas of groundwater depression cones, exploring pilot measures of farmland rotation and fallowing mechanisms, developing scientific methods for the conjunctive use of surface water and groundwater and various types of unconventional water sources, strictly controlling groundwater exploitation, improving the national groundwater monitoring system, and comprehensively managing groundwater over-exploitation areas.

All these documents prove that groundwater depletion has been paid great attention to on the central government level since the beginning of the 2010s. This has resulted in the deployment of the pilot program "Comprehensive Control of Groundwater Overdraft in North China Plain" in Hebei Province in 2014, which has been supported by the Ministries of Finance, of Water Resources, of Agriculture and Rural Areas, and of Land Resources (now Natural Resources). The pilot program started in four prefectures (including Handan) in Hebei Province, covering 49 counties (including Guantao County in Handan Prefecture).

2.2 Groundwater Over-Pumping Control Measures in Hebei Province

Different governmental departments in Hebei Province contributed different measures to reach the goal of groundwater overdraft control. From 2014 to 2017, Hebei Province yearly issued a report on "Integrated Pilot Planning for Governing Groundwater Overdraft". The measures used by various governmental departments are listed in Table 2.1. They cover all sectors involved, including Agriculture, Forestry and Water Resources, and can be categorized into two classes: demand side measures and supply side measures. Among the demand-side measures agriculture contributed mainly with subsidized fallowing of winter wheat, while forestry gave incentives to convert winter crops to water saving non-food crops such as trees. The water resources sector contributed by water saving irrigation technology and a water price reform system. The main activity of the water resources sector was, however, on the supply side, consisting of massive imports of surface water from the South (Yellow River and Yangtze River) to replace groundwater pumping and to increase aquifer recharge.

The effects and challenges of all new measures listed in Table 2.1 are assessed in the following sections.

2.2.1 Seasonal Land Fallowing

The Seasonal Land Fallowing Program (SLFP) was introduced as one of the important measures for groundwater overdraft control in Hebei. The main fallowing crop is winter wheat, which needs irrigation during its growth season from October through

Table 2.1 List of measures used in Hebei Province for groundwater overdraft control from 2014 to 2017

Sector/Department		Measures
Agriculture		• Seasonal land fallowing (new) • Fertigation (traditional)
Forestry		• Substitution of non-food crops for grain crops (new)
Water resources	Water conservation	• Replacing groundwater by surface water (new) • Irrigation water saving technologies (traditional)
	Agricultural water price reform	• Buy-back of water rights (new) • Increase of prices and provision of subsidies (new) • Tiered scheme of water fees (new)

Note New measures are the ones not used before 2014. Traditional measures have been used before 2014

Table 2.2 Subsidized fallowed area and groundwater saved through subsidized fallowing in Handan and Guantao from 2014 to 2020

Year	Subsidized fallowing area in Handan (mu)	Subsidized fallowing area in Guantao (mu)	Reduction in groundwater pumping in Guantao (Mio. m³)
2014	50,000	5,000	0.80–0.90
2015	73,500	15,000	2.40–2.70
2016	183,500	35,000	5.60–6.30
2017	226,656	37,000	5.92–6.66
2018	305,160	42,000	6.72–7.56
2019	395,400	62,000	9.92–11.16
2020	395,400	62,000	9.92–11.16

Note The range of reduction in groundwater pumping in Guantao is calculated with the irrigation norm for winter wheat as lower estimate (160m³/mu/year) and the claimed water saving from the seasonal land fallowing program in Hebei Province as upper estimate (180 m³/mu/year)

May. Seasonal land fallowing has an immediate effect on groundwater table recovery. Water which is not pumped remains in the aquifer. The annual water saving claimed for fallowing winter wheat is up to 180 m³/mu (1 mu = 1/15 ha).

Seasonal land fallowing has been implemented in Handan Prefecture including Guantao County. The subsidized fallowed area in Guantao from 2014 on is listed in Table 2.2. The subsidized winter wheat fallowing in Guantao implies a reduction of groundwater pumping by up to 10 Mio. m³ per year under current funding. (Note that the net groundwater saving compared to winter wheat planting is only 80% of that figure when considering the irrigation backflow). It is a substantial contribution towards elimination of Guantao's groundwater gap. The fallowed area determined by remote sensing is virtually identical to the official figures (see Sect. 4.2.4).

Regardless of the success of seasonal land fallowing (mainly winter wheat fallowing) regarding groundwater saving, there is no clear prospect for a sustainable funding source, unless the central finance authority can provide funds in the future. To eliminate over-pumping, the measure should cover a larger area, which could contradict the country's grain security policy. The farmers' feedback and other challenges of the Seasonal Land Fallowing Program are described and discussed in more detail in Chap. 3.

2.2.2 Substitution of Non-food Crops for Grain Crops

Substitution of winter wheat by non-food crops was implemented mainly through subsidizing the planting of drought resistant tree species, with intercropping of forage grass, medicinal herbs, and other drought-tolerant crops instead of grain crops. Main tree species are fast growing poplar, Chinese ash, locust tree, and walnut tree. The subsidy standard is 1500 CNY/mu in the first year and 750 CNY/mu/year from the second to the fifth year (altogether 4500 CNY/mu in five years). Farmers do not need any more subsidy after the five-year period.

Replacing grain crops with trees can ideally save water for irrigation of winter crops, however the newly planted trees also need irrigation for the first few years before their roots have grown deep enough to completely rely on precipitation and deep soil water. Farmers can expect higher income through the subsidy in the first few years, but the income afterwards depends both on the market price of the trees and the management and technology for pest control. In some places it was reported that farmers planted fruit trees (for example pear trees), which in contradiction to the measure's purpose require intensive irrigation. Tree planting has been implemented in Handan and Guantao, but only on a very small scale. Its contribution to closing of the water gap is negligible up to now.

2.2.3 Replacing Groundwater by Surface Water

Since 2014 the Central Government has funded numerous engineering projects in all pilot prefectures in Hebei Province, involving dredging of river channels, renovation of the canal system and digging pits or ponds to improve the surface water supply system with storage facilities. All these measures aimed at increasing the fraction of the cropping area, which can be irrigated by surface water.

In Hebei Province, surface water is imported mainly through three projects, the SNWT, the Yellow River diversion and the Wei River diversion (Table 2.3). The surface water imported through the SNWT Project has so far been used to replace groundwater abstraction for households and industry in urban areas within the project's reach. In this way, over 80% of the urban groundwater abstraction has been eliminated by the end of 2020. Moreover, in areas with groundwater of high fluoride content and suitable local water supply networks, the imported surface water has been used to replace the extraction of deep confined groundwater and provide

Table 2.3 Effect of imported surface water on groundwater over-pumping control in 2014–2016

Project	Year of implementation	Imported surface water (Mio. m^3)	Number of wells closed in urban/rural areas	Groundwater replaced in rural areas (Mio. m^3)	Affected population/farmland in rural area
SNWT	2014–2016	511	4964	197	4.34 Mio. Capita
Yellow River diversion	2014–2016	784	6898	360	3.28 Mio. Mu
Wei River diversion	2014–2016	157	1649	71	0.7 Mio. Mu

safe drinking water for 4.3 million people in 50 counties of Cangzhou, Hengshui, Xingtai, Handan and Langfang. The Yellow River diversion project and its water supply network could currently play a dual role of both importing and storing water, as a series of water storage projects such as canals, pools and ponds have been constructed simultaneously with the diversion works. From 2014 to 2016, in total 784 Mio. m^3 of water from the Yellow River was transferred to Hebei and close to 7000 irrigation wells were shut down. The Wei River diversion project, which is traditionally supplying water for irrigation purposes, serves its task more efficiently through the updating of irrigation facilities in some small-scale irrigation districts. From 2014 to 2016, the actual amount of irrigation water applied increased by more than 150 Mio. m^3 without increasing the total amount of imported water. In the same period, close to 1700 irrigation wells were shut down. According to the report of the pilot program "Comprehensive Control of Groundwater Overdraft in Hebei Province" (MWR/GIWP 2019), the imports of surface water contributed more than 40% of the total decrease of groundwater abstraction from 2014 till the end of 2018.

Through the import of surface water, groundwater overexploitation in urban areas of Hebei Province has been brought under control. In places where surface water is provided for irrigation, groundwater extraction in the agricultural sector has been significantly reduced. According to an assessment of the water import project conducted by a third party, from 2014 to 2016 the decline rate of groundwater levels in the pilot area has been decreasing under unchanged precipitation conditions (MWR/IHWR 2014). Ecological and environmental problems such as land subsidence and downward trend of the interface between saline water and fresh water in some areas have been alleviated. No further deterioration of the ecological environment has been observed.

The measure has an immediate and direct effect on saving groundwater and in addition increases net recharge through irrigation backflow. It also exhibits a high level of acceptance by farmers. The unit surface water price being lower than the pumping cost of the same amount of groundwater, they can cover their irrigation water needs at lower price.

However, the primary constraint is that surface water is insufficient, and supply is not reliable, neither spatially nor temporally. It often fails to arrive at the time of need. This was the reason why groundwater pumping had become so popular in the first place. The implementation of this measure mainly focused on the construction of canals, infiltration ponds, and pumping stations, while neglecting management including timing, which is the main prerequisite for distributing surface water efficiently. Although amounts imported to Guantao after 2014 look impressive (Table 2.4), their use has been extremely inefficient due to lacking infrastructure for distribution and storage.

Table 2.4 Surface water imported to Handan and Guantao, including water for both domestic and agricultural use

Year	Surface water imported to Handan (Mio. m^3)	Surface water imported to Guantao (Mio. m^3)
2014	643.25	31.87
2015	689.54	36.00
2016	658.32	36.00
2017	794.22	39.00
2018	852.14	43.00
2019	800.00	59.00
2020	862.41	43.10

Source Handan General Management Office of Water Resources, Handan Water Resources Bulletin 2014–2020. Note that these figures contain surface water imported from both inside and outside the NCP. They cannot be compared to figures in Table 2.3. The Guantao figures were used in the modelling in Chap. 4

2.2.4 Buy-Back of Water Rights

Buyback of water rights was a measure promoted by the Ministry of Water Resources and tested in Cheng'an County in Handan Prefecture, starting in 2016. It integrated 10 villages in a monitoring platform. In 2017, the network was extended to over 70 villages. The ideal scheme for this system makes the water user association (WUA) buy back unused water rights from farmers at higher price, to subsequently sell them to other WUAs either through government or directly. The measure, which is illustrated in Fig. 2.1 was intended to enhance groundwater trading and establish a groundwater market. In reality, however, there was no WUA buying water from

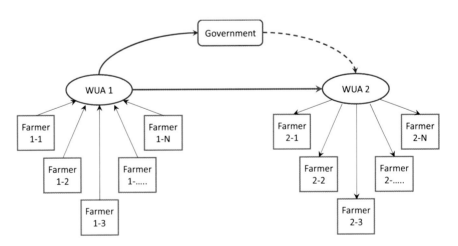

Fig. 2.1 Water right transfer model planned in Cheng'an

another WUA. The government had to buy back the water rights at CNY 0.2 per m^3. Farmers may trade informally among themselves, but they do not trade through the network in order to avoid the higher water price.

A positive outcome is that farmers' consciousness of water saving has been improved, once they got to know that saved water can be sold. Water sold to a WUA is not used as it is not traded to another WUA, which implies real water saving. Water trading, on the other hand, only increases water value. While it may incentivize individual farmers to save water on their own field, it will not reduce regional pumping as the traded water will also be used, only by users who can pay more.

The system in Cheng'an County had to be abandoned after a couple of years of trial due to lack of adequate and stable government repurchasing funds. Only 100,000 CNY available from the provincial government in 2016, were not enough to satisfy the sellers' demand in 2017, which was so high that government had to suspend the buy-back. A large number of informal water rights transactions exists within farm neighborhoods, who do not bother to go through the trouble of using the formal water rights market for the small amount of water rights they can trade. There is no high-value-added agriculture or local industry water user who can buy additional water rights at higher price from the government. Such users, however, have proven to be an important prerequisite for a successful water market as established in Northwestern China.

2.2.5 *"Increase Price and Provide Subsidy"*

As part of the agricultural water pricing reform actions, "Increase price and provide subsidy" has been practiced in 14 counties in Hengshui prefecture, but not in Handan and the other two pilot prefectures. The mechanism of the reform measure is illustrated in Fig. 2.2. The water fee is charged to the farmers every month by their well managers by increasing the electricity price from 0.65 to 0.95 CNY/kWh. The price increase of 0.30 CNY/kWh together with an additional subsidy of 0.15 CNY/kWh from local government is collected in an account managed by the Water User Association of the village. These funds are paid back to farmers according to their water saving performance twice a year in order to give an incentive for water saving while guaranteeing that farmers' income is not affected by the additional water fee.

The field research in one of the pilot districts in Hengshui, Taocheng District, showed that the local farmers' groundwater use for irrigating wheat and cotton decreased by 21% each. However, if no subsidies were granted, half of the region's farmers would lose money due to the increased electricity/water price. However, with the subsidy most farmers in the pilot villages were able to even earn some extra money (Wang et al. 2016). In another pilot site, Anping County, the scheme was less successful: 33 villages conducted the reform in 2017 and many of them abandoned the scheme in 2018. There was no obvious water-saving effect. The local government just wanted to accomplish the task given by the higher government level of making a reform, without the determination to have a real reform.

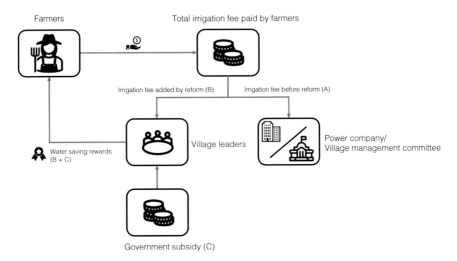

Fig. 2.2 Mechanism of "Increase Price and Provide Subsidy" pilot reform for groundwater irrigation in Taocheng District in Hebei Province

In the implementation process, the inclusion of the water fee in the electricity fee did not make the farmers feel that there is a fee for water. Regardless of the provincial government's request for inter-departmental cooperation, the electricity supplier was not willing to cooperate in the fee collection due to a conflict of interest. For them the scheme meant less sales of electricity and more administrative work. Due to the lack of a data sharing mechanism, the Department of Water Resources had to put considerable efforts into installing water meters and keeping water records, which induced heavy costs for the whole accounting process.

2.2.6 Tiered Scheme of Water Fees

The tiered scheme of water fees is also called the "Three lines and four ladder steps" method for determining the water fee (Fig. 2.3). The scheme is formulated, but calculation of water resources tax and fee according to use relies on installation of measurement facilities and metering. To meter millions of primitive irrigation wells in the NCP with smart water meters is almost an impossible task. It requires huge investments not only for the meters themselves but also for reconstructing wells and their piping and providing appropriate housing to protect the meters. In addition, such a system produces high maintenance cost. In 2017 Hebei Province adopted our suggestion for metering pumped volumes by proxy through electricity consumption. As every well has an electricity meter, this solved the difficulty of metering the large number of small, primitive wells. However, the fee collection could not be implemented up to now, neither in Guantao nor in any other pilot region in Hebei

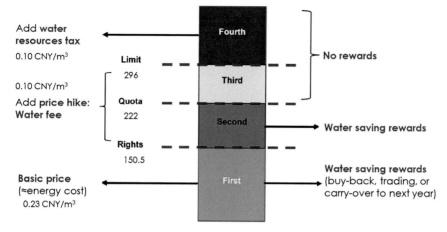

Fig. 2.3 Illustration of "Three lines and four ladder steps" fee collection system. The values of water limit, quota and rights are for Guantao

Province, due to the scheme's complex structure, the difficulty of collecting small amounts of money from millions of individual farmers and last not least the farmers' opposition to any fees.

The proposed agricultural water-pricing scheme is illustrated in Fig. 2.3. The basic price contains the electricity fee and an eventual management fee for the well manager. The water resources tax is 0.1 CNY/m^3, the water fee for usage between water right and water limit (called price hike in the figure) is 0.1–0.2 CNY/m^3, the water saving reward for grain crops staying below the quota is 0.2–0.3 CNY/m^3. If a farmer stays even below the water right, he/she is eligible for a water saving reward e.g. in the form of buy-back at higher price or carry-over to the next season. There is not yet a unified standard among counties. The values of the three lines depend on the local hydrogeological situation and are expressed in units of m^3/mu/year. In Guantao the figures are 150.5 m^3/mu/year for the **water right**, 222 m^3/mu/yr for the **quota** and 296 m^3/mu/year for the **limit**. The difference between a fee and a tax is where it ends up: The fee stays in the county while the tax goes into the state treasury.

Unlike the situation in Hengshui, Guantao Department of Water Resources managed to establish a close and friendly cooperation with the Electricity and Power Supply Company (EPSC) of Guantao. Electricity data were shared for the calculation of pumped volumes. The data includes monthly data of each district and yearly data of each irrigation well. Together with the conversion factors described in Chap. 4, this enabled us for the first time to estimate with good accuracy how much groundwater was pumped in each district and well.

The tax scheme is based on water use per area. In principle, it is possible to calculate the pumped groundwater volume per unit area for each village and each family from the land area and the electricity consumption recorded by village electricians and well managers. It is not possible to do so with the easily accessible electricity data for each well, as the area irrigated by a well can change considerably with the season and from year to year. The flexibility is made possible by the underground piping infrastructure created since the World-Bank-financed GEF project.

DWR is responsible for calculating each well's specific water use. Since 2017, they report every well surpassing the limit to the tax department, which in turn is responsible for collecting the tax. Although water tax has been due for a number of wells, the tax department did not collect the tax, the amount generated in the irrigation sector being minor compared to the cost of organizing the tax collection. Water tax has, however, been collected from industrial users since 2016.

Water fees for wells surpassing the quota have been calculated by DWR, but fees were not collected due to farmers' opposition. Moreover, present fees for water use above the quota are too low to make an impact. On average, no township exceeds the limit of 294 m^3/mu/year relevant for taxing. Average water use of the county as a whole does not exceed the quota of 222 m^3/mu/year. In townships with exceedance of the quota, the amounts are between 20 and 50 m^3/mu/year, the fee for which would amount to insignificant 2–5 CNY/mu/yr. Only greenhouse farmers will have higher water fees, which they, however, can easily pay due to higher profits from their cultures. It has been suggested that present fees should be low to introduce the system smoothly. Once it is in place, fees can be increased. Only if fees lead to water saving, they are justified. If they have no effect on water use, they should either be increased to a level inducing water saving or be abandoned.

The option of automatic smart water metering was tried out in an experiment, collecting the water fee directly via smart cards, but the investment and maintenance costs were both exceedingly high compared to the fees collected. Most meters were broken within a couple of years of installation.

The fee according to the current scheme can only be calculated at the end of the irrigation season, involving all surface and groundwater use. By that time, mistakes of the past season can no longer be corrected. A price solution may have an immediate effect on farmers' irrigation practice. In addition, a price solution does not require the knowledge of the area irrigated by each well. We therefore proposed a direct collection of water fees together with and proportional to the electricity fees. This would make an extra collection system for water fees obsolete. Up to now, Guantao EPSC is not willing to provide this service. The water fee collection system has promoted the consciousness among farmers that water is a valuable resource, but lacking the implementation in practice, it does not unfold its potential to promote water saving.

When choosing the water price solution, the water price (added to present electricity cost) should be at a level equal to farmers' marginal earnings from agricultural production. For example, if marginal earnings in planting wheat amount to 100 CNY/mu and 160 m^3/mu/year are required for irrigation that means the water price should be at least 0.6 CNY/m^3. At that level, farmers on the margin would abandon wheat planting.

2.2.7 Import of Surface Water Versus Water Saving and Change of Cropping Structure

In 2019 the State Council approved the "Integrated action plan for groundwater overdraft control in North China Plain" (or in short form: Action Plan 2019). This plan lists 18 main control actions grouped in four categories of instruments. It is hoped that these allow to reach the target of reducing annual groundwater exploitation by 2.57 Bio. m^3/year (or about 70% of the remaining annual groundwater over-exploitation) in Beijing, Tianjin, and Hebei (BTH) region by 2022. Note that in this action plan the already existing annual surface water imports to BTH region are not included. In 2018 they reached a total of 4.5 Bio. m^3/year, basically eliminating the over-pumping in Beijing and Tianjin.

The Action Plan 2019 lists the measures to be implemented in BTH region by 2022 to further reduce groundwater pumping. They include water saving, change in cropping structure, and replacement of groundwater by surface water imported from the SNWT, the Yellow River and other local water resources. The targeted reductions of groundwater pumping each measure can achieve are listed in Table 2.5. The long-term goal is to eliminate over-pumping completely by 2035.

The main measures recommended in the Action Plan 2019 are the ones, which have proven to be effective in the pilot projects. Figure 2.4 lists the implementation progress in 2019 in terms of the main measures of the Action Plan 2019. To appreciate the tasks, a few comments are helpful. A reduction of pumping by 0.24 Bio. m^3/year through water saving measures on 4.17 Mio. mu of farmland is ambitious. It averages to a saving of 57.6 m^3/mu/year, which is equivalent to the amount of a single irrigation event and probably overestimates the water saving capacity for grain crops. For greenhouses, savings of this order of magnitude are feasible through drip irrigation. Cropping structure change is mainly implemented through winter wheat fallowing and could save 0.60 Bio. m^3/year through fallowing of 4.85 Mio. mu (or 0.32 Mio. ha) by 2022. This amounts to 13.6% of the wheat planting area of Hebei (2.36 Mio. ha in 2018 according to China's Statistical Yearbook) and might be difficult to achieve. Importing surface water seems less difficult to accomplish. Using it to replace groundwater in irrigation still poses problems: The farmland infrastructure, which is mainly based on groundwater irrigation, is not yet capable to receive such large amounts of surface water. The canal system is insufficient as are storage facilities, which would allow to match the timing of surface water imports to the irrigation calendar. This is also the reason why a large percentage of surface water imports accomplished by the end of 2019 (3.49 Bio. m^3 out of 8 Bio. m^3 shown in Fig. 2.4) could only be used for artificial groundwater recharge through rivers and lakes. Note that the progress in the first two measures shown in Fig. 2.4 is expressed in area equipped. While the bars for the accomplished task look longer for water saving than for change of cropping structure, the corresponding amounts of water saved are clearly larger for the second item.

The groundwater saving capacity added through the newly implemented measures in BTH area in 2019 is shown in Fig. 2.5. In total, only an additional 0.67 Bio. m^3 of

Table 2.5 Tasks of groundwater over-pumping control in BTH area to be reached by 2022 (*Source* Action Plan 2019)

Units (10⁹ m³/year)	Current over-pumping	Reduce pumping through		Replace groundwater though imports of water from				Total pumping reduction	Residual
		Water saving	Cropping structure change	SNWT eastern route	SNWT central route	Yellow River	Local water sources		
Total	3.47	0.24	0.6	1.26	0.17	0	0.3	2.57	0.9
Beijing	–	0	0	0	–	–	–	–	–
Tianjin	0.16	–	–	0.01	0.02	–	0.04	0.07	0.09
Hebei	3.31	0.24	0.6	1.25	0.15	–	0.26	2.5	0.81

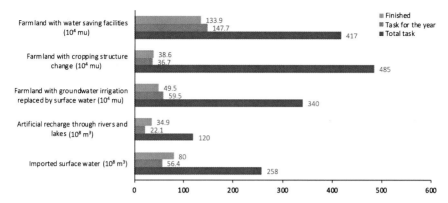

Fig. 2.4 Main measures listed in Action Plan 2019 for BTH area: Implementation progress of year 2019, in comparison to task figures for 2019 as well as the whole implementation period until 2022. Data source: (MWR GIWP and Hai River Comission 2020)

Fig. 2.5 Groundwater saving achieved by different measures in 2019 (Units: 10^8 m^3). *Source* (MWR GIWP and Hai River commission 2020)

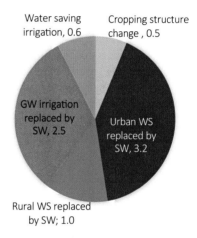

imported surface water could be used to replace groundwater pumping in comparison to a total amount of 5.6 Bio. m^3 imported to the region. Within this small replacement, more than 60% was used to replace groundwater supply for domestic use in both rural and urban regions. Replacing groundwater by surface water in irrigation contributed less than one third to the total increase in groundwater saving for the year. Out of 250 Mio. m^3 only 70 Mio. m^3 was diverted to the newly equipped farmland (495′000 mu), while the rest was diverted to farmland, which had already been equipped with surface water irrigation infrastructure before 2019, but never received water.

Cropping structure change (mainly through fallowing of winter wheat) contributed about 50 Mio. m^3 of groundwater saving, water saving irrigation another 60 Mio. m^3. Both measures seem to contribute little, but in comparison to the real increase of surface water irrigation capacity (70 Mio. m^3), all three measures are important in achieving the goal of reducing groundwater pumping, especially when considering that farmland with surface water irrigation capacity might still use groundwater when surface water does not arrive on time. Increasing the efficiency of surface water use in replacing groundwater in irrigation should be the main focus in the future.

In 2019 the actual amount of surface water imported to the BTH region was 8 Bio. m^3 in total. More than 40% were used for artificial recharge in rivers and lakes to promote ecological rehabilitation, in exceedance of planned figures, as the ability of agricultural land to receive surface water for irrigation was too small. In this situation, MAR is a convenient way of making use of excess surface water not used for irrigation. Two examples are shown here.

In Guantao County, two infiltration basins were built (Fig. 2.6). They were filled once or twice a year with off-season surface water from the Yellow River. Seepage conditions were not ideal in Guantao's rather clayey soils. Still, according to our measurements they allowed about half of the applied water to infiltrate, while the other half evaporated (Mérillat 2016). The volume of a full basin of 35,900 m^2 area was 110,000 m^3 per filling. Such ponds can at the same time serve as short term water storage, holding surface water until it is needed, to increase the efficiency of its utilization.

Better results were achieved by discharging excess water from the SNWT into some of Hebei's dry riverbeds, leading to a significant groundwater table rise along the river courses. One example is shown in Fig. 2.7. Water from the SNWT was diverted into the bed of the usually dry Fuyang River in Hebei via a weir in the location, where the Central Route of the SNWT crosses the river. The observation wells in the vicinity of the river showed a rise of the groundwater table of 0.5 m to more than 5 m. Infiltration of South-North water in the outskirts of Beijing has also been successful in raising groundwater levels (Long et al. 2020). Of course, due

Fig. 2.6 Infiltration basin in Guantao County with water depth gauge (left) and floating evaporation pan (upper right) to measure evaporative losses. A satellite image (bottom right) shows the extent of the infiltration pond, which is called "Moon Lake"

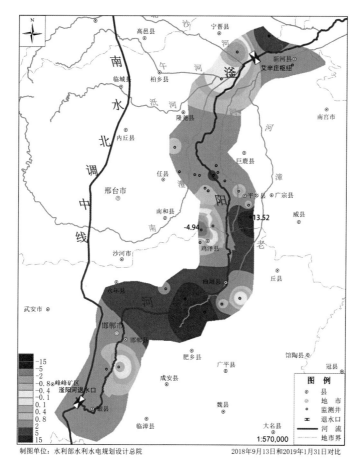

Fig. 2.7 Infiltration in river bed: Example Fuyang River, Hebei. Reaction of groundwater table to water release from South-North Transfer Central Route. Three blue tones in increasing intensity: water table rise, up to 2, 5 and 15 m. *Source* MWR GIWP (2019)

to the high cost of water imports, their use for MAR can only be the option of last resort.

Using surface water for irrigation saves the energy required for pumping groundwater and allows to rehabilitate not only land adjacent to lakes and rivers, but areas in larger need, such as the most severe overdraft zones. In addition, it contributes to groundwater recharge by irrigation backflow. Improving conjunctive allocation of various water resources for the region should be the focus in the coming years to make sure the imported surface water replaces groundwater in irrigation to its maximum potential.

2.3 Governance Structure in the Water Sector

Water governance comprises all processes of governing the supply and use of water by relevant institutions throughout society by means of laws, norms, and regulations. As everywhere in the world, it is a multi-stakeholder affair in China. There are three main institutional systems involved: The Ministry of Water Resources with its substructures on provincial and county levels, the Ministry of Agriculture and Rural Affairs with similar substructures and the State Grid Corporation of China with its branches on provincial and county level. Another but lesser player is the Ministry of Natural Resources, whose competences are mainly in groundwater monitoring and resource capacity assessment rather than groundwater governance. The Ministry of Agriculture and Rural Affairs is the national agricultural policy maker, and no agricultural measures can be taken without its consent. Agriculture as the main water user is responsible for 60–80% of total groundwater pumping in the NCP. With the cropping structure governed through its subsidies, the Ministry of Agriculture has the largest influence on limiting over-pumping. At the same time, it has the—conflicting—national task of upholding food security. The Ministry of Water Resources has the decisive power in supplying water infrastructure and diverting surface waters, which explains their preferred strategy of water transfers. The electric power utilities as the power supplier for pumping have a very comprehensive and fully functional metering and fee collection network for electricity consumption. Since the electricity suppliers certainly opt for selling as much electric energy as possible, pumping control in principle harms their interest. Involving electricity suppliers is crucial in designing pumping metering, fee collection and control strategies for groundwater usage.

The ideal case would be a smooth cooperation among the three governing structures serving the farmers as private stakeholders, who in the end will have to cooperate in the implementation of policies. In practice, the existing cooperation is often hampered by the struggle for competences, power, and funds among the three governing structures. While all governance in China seems top-down, it cannot ignore the wishes of the governed, in this case the farmers, who can render any policy ineffective by refusing to cooperate.

2.3.1 Governmental Stakeholders in Water Sector

On the central government level, Ministry of Water Resources and Ministry of Agriculture and Rural Affairs are both working under the State Council. Departments of Water Resources and Agriculture and Rural Affairs are governmental bodies, which receive administrative orders from the local government (provincial, prefecture and county levels), and technical instructions from their respective ministries, as shown in Fig. 2.8. The heads of departments are normally employed by the local governments, while the technical staff working in different departments are hired through

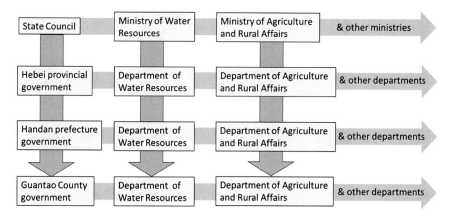

Fig. 2.8 Illustration of matrix structure of governmental administration in China involved in groundwater over-pumping control in Guantao County

government funds, or partially through project funds procured either from the local governments or from ministries.

The program "Integrated Pilot Planning for Governing Groundwater Overdraft" was initiated by the central government in Beijing with funds given to Hebei Province directly. Four prefectures in Hebei Province (including Handan) were chosen as pilot sites, different departments on prefecture level took up the task to implement the measures and distribute the work among their respective county level departments. The ministries involved supported the national initiative of over-pumping control by providing planning guidelines and recommending and promoting various measures.

2.3.2 Stakeholders in the Electricity Sector Related to Irrigation

In China electricity supply and its distribution network belonged to the State Power Corporation of China, a state enterprise founded in 1997 and dismantled in 2002 due to the institutional reform of the electric power sector. Since then, the power supply grid in the NCP belongs to the State Grid Corporation of China, a state-owned enterprise. The company has sub-branches in Hebei Province, Handan Prefecture and Guantao County.

In Guantao County, electric energy is supplied and managed by Guantao Electric Power Supply Company (EPSC), which is affiliated with the State Grid Corporation of China. Its organizational and electricity metering systems are shown in Fig. 2.9.

The EPSC is in charge of eleven Electric Power Supply Agencies (EPSA) at district level. Each EPSA manages the electric power supply of about twenty villages. In each village, one or two electricians employed by EPSA are responsible for power infrastructure maintenance. The village electricians are also responsible for collecting

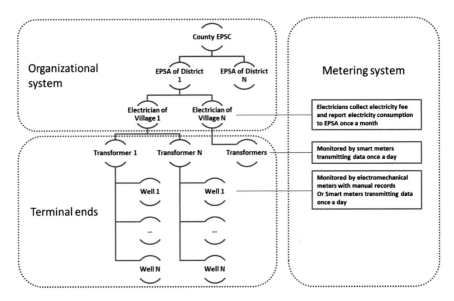

Fig. 2.9 Organizational system and terminal ends of the electric power distribution system (left boxes) and the metering system (right box)

electricity fees from the end users. In the case of irrigation, the end-users are the well managers who are owners of a well or take care of a well, owned by several households. For a well shared by several farmers, a written chronological log of pump operations is kept, recording the readings of the electricity meter when the pump is turned on and off. The electricity fees are collected by the well manager on a monthly basis, according to the energy consumption records of individual farmers. Based on the existing electricity metering system, groundwater pumping can be monitored at different levels. Since 2018, all wells are equipped with smart electricity meters, which transmit daily electricity consumption to the utility.

2.3.3 Stakeholders in the Water Sector of Guantao County

The Department of Water Resources in Guantao County is the main governmental stakeholder playing a major role in the pilot program for over-pumping control. It is responsible for implementing the engineering measures of replacing groundwater by importing surface water, through construction of canals, infiltration ponds, and surface water pumping stations. The department is also responsible for promoting water saving by providing subsidies for water saving irrigation equipment and carrying out the agricultural water pricing reform.

On one hand, the Department of Water Resources in Guantao made a huge effort to set up metering and fee collection for imported surface water. However, for groundwater pumping control, which involves individual farm households, they have to rely on the village administration. In 2007, a World Bank project tried to set up Water User Associations (WUA) in each village. Due to China's strict control of the number of non-governmental organizations within a county, eventually a WUA was set up in Guantao as one organization for the whole county, each village's WUA being a sub-branch of it. The village WUA therefore normally only comprises the village leaders and part of the village administration, receiving governmental orders instead of being the bottom-up initiative, as which it is known outside of China.

In recent years after our strong recommendation and also due to the fact that groundwater pumping has to be metered through pumping electricity, most of the villages have integrated the village electricians as members into the WUAs. As contract employees of Guantao EPSC, village electricians are paid to collect electricity fees from well managers, who themselves receive an additional irrigation service fee from the well users according to the amount of pumping. The governmentally organized WUAs, however, exist in name only. They do not receive any funding, neither from villagers nor from the government. Therefore, their possibly very useful function in collecting water fees cannot be realized so far.

References

Calow RC, Howarth SE, Wang J (2009) Irrigation development and water rights reform in China. Int J Water Resour Dev 25(2):227–248. https://doi.org/10.1080/07900620902868653

Debaere P, Richter B, Davis K, Duvall M, Gephart, J, O'Bannon, et al. (2014). Water markets as a response to scarcity. Water Policy 16(4):625–649.https://doi.org/10.2166/wp.2014.165

Etchells T, Malano H, McMahon T (2006) Overcoming third party effects from water trading in the Murray-Darling Basin. Water Policy 8:69–80. https://doi.org/10.2166/wp.2006.0005

Frija A, Dhehibi B, Chebil A, Villholth KG (2015) Performance evaluation of groundwater management instruments: The case of irrigation sector in Tunisia. Groundw Sustain Dev 1:23–32. https://doi.org/10.1016/j.gsd.2015.12.001

Heaney A, Hafi A, Beare S, Wang J (2005) Water reallocation in Northern China: towards more-formal markets for water. In: Willet IR, Gao Z (eds) Agricultural water management in China. Australian Center for International Agricultural Research, pp 130–141

Howe C, Schurmeier D, Shaw W (1986) Innovative approaches to water allocation: the potential for water markets. Water Resour Res 22(4):439–445. https://doi.org/10.1029/WR022i004p00439

Huang Q, Wang J, Rozelle S, Polasky S, Liu Y (2013) The effects of well management and the nature of the aquifer on groundwater resources. Am J Agr Econ 95(1):94–116. https://doi.org/10.1093/ajae/aas076

Johansson R, Tsur Y, Roe T, Doukkali R, Dinar A (2002) Pricing irrigation water: a review of theory and practice. Water Policy 4:173–199. https://doi.org/10.1016/S1366-7017(02)00026-0

Kemper KE (2007) Instruments and institutions for groundwater management. In: Giordano M, Villholth KG (eds) The agricultural groundwater revolution: opportunities and threats to development. CABI, Wallingford, pp 153–172

Lewis D, Zheng H (2018) How could water markets like Australia's work in China? Int J Water Resour Manag 34(3):1–21. https://doi.org/10.1080/07900627.2018.1457514

Long D, Yang W, Scanlon BR, Zhao J, Liu D, Burek P, Pan Y, You L, Wada Y (2020) South-to-North Water Diversion stabilizing Beijing's groundwater levels. Nat Commun 11:3665https://doi.org/10.1038/s41467-020-17428-6

Mérillat A (2016). Estimation of the infiltration rate from an artificial lake in Guantao County, China. Master thesis, Institute of Environmental Engineering, ETH Zurich.

Moore S (2015). The development of water markets in China: progress, peril, and prospects. Water Policy, 17(2). https://doi.org/10.2166/wp.2014.063

MWR (2010). Technical Code for Water Wells. Published by Ministry of Housing and Urban-Rural Development and State Administration for Quality Supervision and Inspection and Quarantine.

MWR/IWHR (2014). The third-party evaluation report on the pilot project of comprehensive groundwater overexploitation control in Hebei Province. Report of IWHR (in Chinese).

MWR/GIWP (2019). Comprehensive Control of Groundwater Overdraft in Hebei Province. Report of GIWP (in Chinese).

MWR GIWP (2019). Final evaluation report on pilot project of artificial recharge through rivers and lakes of integrated action plan for groundwater overdraft control in North China Plain. Report of the GIWP (in Chinese).

MWR GIWP and Hai River Commission (2020). The summary and evaluation report of integrated action plan for groundwater overdraft control in North China Plain in 2019. Report of the GIWP (in Chinese).

Shen D (2015) Groundwater management in China. Water Policy 17:61–82. https://doi.org/10.2166/wp.2014.135

Speed R (2009) A comparison of water rights systems in China and Australia. Int J Water Resour Dev 25(2):389–405. https://doi.org/10.1080/07900620902868901

State Council (2012). Opinions on the implementation of the strictest water resources management system. Retrieved from http://www.gov.cn/zwgk/2012-02/16/content_2067664.htm

Sun T, Wang J, Huang Q, Li Y (2016) Assessment of water rights and irrigation pricing reforms in Heihe River Basin in China. Water 8(3338). https://doi.org/10.3390/w8080333

Wang J, Zhang L, Huang J (2016) How could we realize a win–win strategy on irrigation price policy? Evaluation of a pilot reform project in Hebei Province, China. J Hydrol 539:379–391. https://doi.org/10.1016/j.jhydrol.2016.05.036

Wang J, Li Y, Huang J, Yan T, Sun T (2017) Growing water scarcity, food security and government responses in China. Glob Food Secur - Agric Policy Environ 14:9–17. https://doi.org/10.1016/j.gfs.2017.01.003

Wang J, Jiang Y, Wang H, Huang Q, Deng H (2020) Groundwater irrigation and management in northern China: status, trends, and challenges. Int J Water Resour Dev 36(4):670–696. https://doi.org/10.1080/07900627.2019.1584094

Wang J, Zhu Y, Sun T, Huang J, Zhang L, Guan B, Huang Q (2020) Forty years of irrigation development and reform in China. Aust J Agric Resour Econ 64(1):126–149. https://doi.org/10.1111/1467-8489.12334

Yu X, Geng Y, Heck P, Xue B (2015) A review of China's water management. Sustainability 7:5773–5792. https://doi.org/10.3390/su7055773

Chapter 3
Cropping Choices and Farmers' Options

Irrigation being the main cause of aquifer depletion, agriculture is the first candidate to contribute to its solution. Options of agricultural planting structure in Beijing-Tianjin-Hebei region are analyzed using various planting scenarios. The analysis shows that when addressing only the region's self-sufficiency in food, planted area can be reduced by 26%, eradicating over-pumping but decreasing farmers' revenue by 50%. No realistic agricultural strategy can eliminate over-pumping in North China Plain without water transfers from the South. Farmers' reaction to policies plays an important role regarding their efficiency. The implementation status and effects of seasonal land fallowing in Hebei Province were evaluated in a field survey of 560 farm households showing that the subsidy is welcome, and farmers are eager to participate in the program. However, more than half of the farmers will go back to winter wheat growing if the subsidy is decreased or discontinued. The groundwater game "Save the Water" was played with farmers in Guantao. Results showed that the farmers are not so much led by profit optimization as by customs and inertia against change. They, however, reacted strongly to the visualized decline of groundwater levels, which indicates that appropriate information may induce behavioral change.

Electronic supplementary material The online version of this chapter (https://doi.org/10.1007/978-981-16-5843-3_3) contains supplementary material, which is available to authorized users.

3.1 Options of Optimizing Crop Structure in Hebei-Beijing-Tianjin Region

3.1.1 Introduction

The Beijing-Tianjin-Hebei (BTH) Plain is the area of most serious groundwater depletion in China (Feng et al. 2013, 2018). In the piedmont plain of the Taihang Mountains the groundwater level dropped most rapidly. It is estimated that its shallow aquifer under present abstractions could be depleted to its physical limit within the next 80 years (Zhang et al. 2016). Hebei Plain has become one of the most vulnerable areas in China and possibly worldwide (Wang et al. 2015).

Food production in the BTH area, the main cause of groundwater overexploitation, increased continuously and today exceeds the region's total food requirements by far. The grain surplus in the region is 49% of the total grain production, if only considering the food grain requirement. It is 9% of the total production if the requirements of the region for feed grain used in the production of meat, eggs and milk are also included. Meanwhile, the surplus amounts of fruit, vegetables, eggs, milk and aquatic products are all more than 50% of the respective production in BTH region (Fig. 3.1). According to the national estimates on water resources and water use, the BTH region overexploited groundwater resources on average by 6.7 Bio. m^3/year between 2005 and 2015. Since 2014, the imports of surface water through the SNWT project have to a large part replaced groundwater use by households and industry, leaving a deficit mainly due to agricultural groundwater use. It is estimated to be about 4.4 Bio. m^3/year (or 65% of the overexploitation between 2005 and 2015), which still makes both water resources use and agricultural production unsustainable.

Irrigation of the intensifying cropping system has become the main cause for serious groundwater depletion. Before the 1970s, Hebei Plain was a dry-land farming area dominated by wheat and millet, without any problem of groundwater overexploitation. Since the 1970s, however, with the improvement of irrigation conditions, the planting system gradually developed in intensity from one harvest per year via

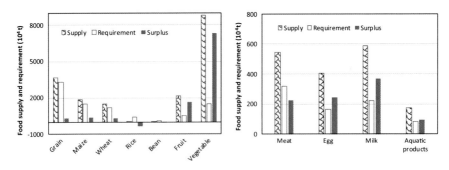

Fig. 3.1 Annual food supply, demand and surplus of households in Beijing-Tianjin-Hebei region (in 10,000 tons) *Source* (Luo 2019)

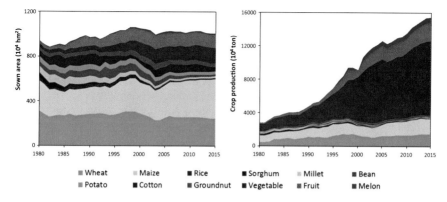

Fig. 3.2 Development of crop production according to sown area (in 10,000 hectares)(left) and annual yield (in 10'000 tons) (right) of crops in Beijing-Tianjin-Hebei region between 1980 and 2015 *Source* (Luo 2019)

three harvests in two years, and four harvests in three years into a high-intensity irrigated agricultural production mode, which is dominated by a winter wheat-summer maize double cropping system with two harvests in a year (Xiao et al. 2013; Wang et al. 2012; Mo et al. 2009).

The planting structure of crops has changed remarkably in the past 35 years. It has become simpler with respect to the crop variety while the crop yield has greatly improved (Fig. 3.2). A planting mode dominated by wheat, maize, fruit, and vegetables evolved. The water consumption of well-watered winter wheat, the main food crop, is approximately 420–430 mm (Shen et al. 2013; Zhang et al. 2011), but the precipitation in its growth season is less than 150 mm (1963–2013), leading to a water deficit of approximately 270–280 mm. For the economic crops, such as vegetables and fruit, the precipitation is also less than their water consumption during the growth period. Green house planting can consume considerably more water per year than any other cropping pattern (including double cropping of winter wheat and summer maize) due to its water intensive vegetable and fruit cultures and its multiple cropping all year round.

As the scale of irrigation expanded, groundwater consumption also increased (Cao et al. 2013). In the period of 1984–2008, 139 Bio. m^3 of groundwater have been consumed by grain production in Hebei Province (Yuan and Shen 2013). In the BTH region, the current irrigated planting area has exceeded the carrying capacity of the region's water resources. Therefore, a reduction in the irrigated planting area has become the key to achieving the required massive water saving on a regional scale. Its implementation is an urgent task.

Reducing or replacing high water consumption crops (such as winter wheat, vegetables, and fruit) is the most efficient way to save water. There are two ways to downsize the irrigation area: one is to de-intensify the cropping system, changing the conventional winter wheat and summer maize double cropping system to a cropping system of for example three harvests in two years (Luo et al. 2018); the other is to optimize the planting structure in terms of economic output under different constraints (Luo 2019).

In this study, the bearing capacity of water resources, the food requirements of BTH region and the self-sufficiency of the region in the main agricultural products were considered as constraints. Four scenarios with different objectives for optimization were defined. Using the Elitist Non-dominated Sorting Genetic Algorithm (NSGA-II), the planting structure was optimized with respect to economic benefits for each of the four objectives, quantifying the optimal planting structure, its water use, and its crop production. The goal was to explore a sustainable planting structure in a tradeoff between water resources requirements and agricultural production, providing a policy basis for the sustainable utilization of water resources and regional food security in the BTH region.

3.1.2 Optimization Scenarios

Four simulation scenarios for optimizing the planting structure were defined. The results are described in terms of water use, land use, and the value of economic output. Apart from the first scenario, which could be reached by 2030 given current trends, the scenarios are not predictions but rather benchmarks, against which actual policy can be calibrated. Therefore, no time scales for their implementation are given.

Scenario I (current development trend): This scenario assumes that the planting structure of crops will evolve according to current trends without being affected by a macro-control and management policy.

Scenario II (self-sufficiency in the main agricultural products): In this scenario, the planting structure of crops will depend on the needs defined by self-sufficiency of BTH region in the main agricultural products (excluding rice).

Scenario III (maximum grain output under the constraints of water resources): In this scenario, the planting structure of crops aims at the maximum grain output scale which can be supported by the water resources available in the area in view of satisfying national grain security requirements.

Scenario IV (coordination of grain crops, cash crops and water needs): In this scenario, the optimized planting structure will be constrained by the scale needed to ensure grain self-sufficiency of the area (total amount of wheat and rice) while maximizing the economic output.

Detailed information on the formulation of the objectives can be found in (Luo 2019; Luo et al. 2021). The following sections will discuss scenario by scenario the results, which are summarized in Table 3.1.

Table 3.1 Optimization scenarios of planting structure in Beijing-Tianjin-Hebei (BTH) region (Luo 2019; Luo et al. 2021)

Scenarios	Current status	Optimization results			
	S0	S1	S2	S3	S4
Total planting area (10^6 ha)	10.17	9.93	7.48	7.35	9.55
Irrigation water consumption (10^9 m^3/year)	12.3	11.3	6.6	5.6	10.5
Economic benefit (10^9 CNY/year)	239.9	239.1	118.6	117.7	213.7

Note: S1—Current trend scenario; S2—regional self-sufficiency in agricultural products; S3—maximum grain output; S4—coordination of grain crops, cash crops and water need

3.1.3 Scenario Analysis of Planting Structure Optimization

Scenario of the current development trend

The first scenario (S1) predicts the future crop planting structure based on current trends in the BTH region. The scenario will result in 2% reduction in the sown area of major crops in the Hebei Plain.

Under the current development trend scenario, the future planting structure will be more water efficient. Major crops in the BTH region can reduce water consumption in total by 1.0 Bio. m^3/year compared to the current total amount (Table 3.1). In addition, the total grain output will increase by 5% with no significant change in economic benefits. The reduction in irrigation water use can reduce groundwater overexploitation to some extent, but it is still far from solving the problem of groundwater overexploitation in the BTH-region.

Scenario of self-sufficiency in the main agricultural products

The maximum potential for water conservation resulting from scenario S2 is an important reference for the formulation of any agricultural and water policy in the BTH region. Under this scenario, the sown area of major crops in the region will be reduced to 7.48 Mio. ha, a reduction of 2.69 Mio. ha (or 26%) compared to the current status. The total grain output will decrease by 11%.

The total amount of water saved by major crops relative to the current planting structure is 5.7 Bio. m^3/year, which may be the maximum potential for water conservation in the BTH region, when only the region's local demand for major food crops is considered.

Comparing to the estimated overexploitation of the BTH region, regional self-sufficiency can solve the problem of groundwater over-abstraction without additional surface water import by the SNWT Project. In this scenario, the economic benefit will decrease by about 50%, while a balance between exploitation and replenishment of groundwater used for farming will be achieved. This scenario can eliminate agricultural groundwater overexploitation.

Scenario of maximum grain output under water resources constraints

The maximum wheat production, which can be supported by regional water resources (S3), is an important basis for determining the maximum scale of agricultural production under a balanced groundwater budget. Based on the national data on current water resources, the mean regional water resources available in the BTH region are 18.8 Bio. m^3/year (average 2005–2015). To maximize wheat production, other major crops with water deficits during the growth period (including vegetables, fruit, cotton, oil crops and potatoes) are constrained by the food consumption needs of the area to ensure that maximum water resources are left for wheat irrigation. Under this scenario, the maximum wheat planting scale that can be supported by regional water resources is 1.48 Mio. ha, equivalent to about 60% of the status quo wheat planting scale.

In this scenario, regional water resources are used as the limit of water consumption to balance exploitation and replenishment of groundwater. The total sown area of crops will decrease by 28% and the total grain output by 13%. The economic output will be reduced, while water saving will balance out the estimated regional overexploitation by agriculture, resulting in a sustainable use of regional agricultural water resources.

Scenario of coordination of grain crops, cash crops and water requirements

The coordinated development of grain crops, cash crops and water use (S4) has long been an issue of great concern in the optimization of planting structures. The scenario optimizes the planting structure by demanding self-sufficiency in food crops for the region (total demand for rice and wheat, where the rice deficit is converted into wheat), groundwater protection, and economic benefits as the critical criteria for the development. The results show a 6% reduction in the crop planting scale relative to the current cropping structure. Wheat, rice, vegetables, and fruit roughly maintain a scale similar to the current planting structure, which can meet the regional requirement for food (wheat and rice), save water and largely secure economic benefits. Therefore, the current planting structure is a relatively reasonable planting structure if one does not take restrictions on water resources into account.

The scenario can save 1.8 Bio. m^3/year of water relative to the current planting structure. It can mitigate the problem of groundwater overexploitation, but the water saved is much less than the estimated 4.4 Bio. m^3/year of overexploitation. Therefore, other water sources are still required for a balanced groundwater budget, while meeting the demand for food crops and economic output.

3.1.4 Conclusion and Discussion

The water deficits of the major crops—wheat, vegetables, and fruit—account for about 90% of the total groundwater consumption in farming. Different planting structure scenarios can alleviate groundwater overexploitation with the amount of water saved ranging from 1.0 Bio. m^3/year to 6.7 Bio m^3/year. The sown area of major crops must be reduced by 2% to 28%, and the scale of winter wheat, a major crop of high groundwater consumption, by 8% to 41% of the current scale. Changes in food production over the scenarios range from 5% increase to 13% decrease, while the reduction in direct economic output in farming ranges from 0.3% to 51%.

Overall, self-sufficiency in the main agricultural products (S2) can meet the regional self-sufficiency in agricultural production and water saving under this scenario is sufficient to achieve a balance between exploitation and replenishment in agricultural water use. Maximum food output under water resource constraints (S3) results in a planting structure scenario with a relatively high degree of sustainability in agricultural water use and a relatively high regional grain self-sufficiency. The two scenarios (S2 and S3) are the preferred optimized planting structures for the BTH region.

The above scenarios provide reference thresholds for restructuring of the planting system to achieve sustainable use of agricultural water resources. They show how big the contribution of agricultural restructuring to sustainable groundwater use can theoretically be and what this means in terms of production and farmers' income. Sustainability in groundwater resources could be reached by agricultural restructuring alone, but it would come with a high price tag regarding farmers' income. In comparison to an income loss on the order of 100 Bio. CNY/year, the cost of additional 2–3 Bio. m^3/year of SNWT-water at a price of 2–3 CNY/m^3 seems affordable. It also must be noted that no scenario can achieve the national goals of grain security **and** a balanced groundwater budget at the same time without additional water imports through the SNWT project.

In practice, the adoption of an optimized crop structure will depend on farmers' behavioral traits as well as yield and market forces and willingness to pay for the water resources. To a certain degree the process can be steered by the state through subsidies, be it on agricultural products, fallowing, water saving technology, or through the water price itself.

3.2 Farmers' Feedback in a Household Survey on Seasonal Land Fallowing

A large-scale field survey was conducted in four prefectures in Southern Hebei Province from April 2018 to September 2019 by the China Center for Agricultural Policy (CCAP) of Peking University. The four prefectures (Cangzhou, Handan, Hengshui, and Xingtai) participated in the Seasonal Land Fallowing Program (SLFP) from the start and account for nearly 90% of the implementation area of Seasonal Land Fallowing (SLF). In addition, these four prefectures are the most serious regions of groundwater overdraft in Hebei Province. Within these four prefectures, 7 counties (Yanshan, Pingxiang, Qinghe, Gucheng, Jizhou, Qiuxian and Guantao, shown in Fig. 3.3) were selected to conduct a field survey. Within each county, two townships

Fig. 3.3 Location of sample
counties for the field survey
in four prefectures
(Cangzhou, Hengshui,
Xingtai and Handan) in
Hebei Province

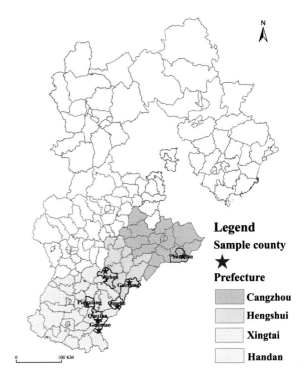

and within each township, two villages were chosen for the survey, one village which
participated in the SLF project and another village which did not. In each village,
20 farm households were randomly selected for conducting a face-to-face interview.
The final sample included 560 households in 28 villages of 14 townships in 7 counties and 4 prefectures. Among the 28 surveyed villages, 14 villages participated in
the SLF and 14 did not. Among the 560 households surveyed, 249 participated and
311 did not.

3.2.1 Effects of Seasonal Land Fallowing

Relatively high targeting efficiency

According to the policy guidelines, pilot sites participating in the SLF should satisfy the following four conditions:

1. They should be located in a groundwater overdraft zone.
2. Irrigation should mainly depend on groundwater.
3. They should grow winter wheat.
4. Land considered for fallowing should cover a coherent area of at least 50 mu (3.3 ha).

The survey's results show that most pilot sites participating in the project satisfied the four requirements. For example, all sample villages participating in the project are located in groundwater overdraft zones in accordance with the definition given by Hebei Province. Among 14 villages, 6 villages are not only located in *General Overdraft Zones of Shallow Groundwater* but in *Serious Overexploitation Zones of Shallow Groundwater.* 8 villages belong to the *Serious Overexploitation Zone of Deep Groundwater.* The definition of the Overdraft Zones can be found in (Hebei Government 2017). In addition, 77% of cultivated land in the sample villages mainly use groundwater for irrigation. Most cultivated land (93%) of the participating villages is concentrated and it is not hard to find a coherent plot larger than 50 mu. About 70–80% of plots within the SLFP were planted with winter wheat before participating. This also means that between 20 and 30% of plots did not plant winter wheat before participating in the SLFP.

Reduction of water use

Our first-hand survey data prove that the SLF can reduce farmers' groundwater use. We compared the changes in water use (Tables 3.2 and 3.3) between SLF households who started to participate in the SLFP in the winter of 2017/2018 and non-SLF households, both before and after the SLFP. We found that the SLF households, who started to participate in SLF in the winter of 2017, reduced their annual water use by 15.7% (735 m^3/ha) in 2019 (Table 3.2). Comparing the same two years, the

Table 3.2 Comparison of annual water use per hectare of non-SLFP households and SLFP households who participated in the SLFP in winter of 2017

	SLFP households			Non-SLFP households		
	2017	2019	Change	2017	2019	Change
Annual agricultural water use per hectare (m^3/ha)	4680	3945	−735	6375	6675	300
Annual agricultural groundwater use per hectare (m^3/ha)	3855	3120	−735	6060	6375	315

Source SLFP survey in 2019

Table 3.3 Comparison of annual water use per hectare of non-SLFP households and SLFP households who participated in the SLFP in the winter of 2018

	SLFP households			Non-SLFP households		
	2018	2019	Change	2018	2019	Change
Annual agricultural water use per hectare (m^3/ha)	4440	3600	−840	6435	6675	240
Annual agricultural groundwater uses per hectare (m^3/ha)	4275	3510	−765	6120	6375	255

Source SLFP survey in 2019

annual water use of non-SLF households increased by about 4.7% (300 m^3/ha) due to differences in precipitation in the years compared. Consequently, it can be estimated that the project led to an annual reduction of **total** water use and **groundwater** use by 20.4 and 24.3% respectively. Similarly, it can be estimated that **total** water use and **groundwater** use of households who started to participate in the SLF in the winter of 2018 (Table 3.3) both decreased by approximately 22%. Hebei Provincial Government claims that SLF can decrease annual irrigation water use by 2700 m^3/ha (180 m^3/mu), while the Action Plan 2019 and the evaluation report used a more realistic number of 120–140 m^3/mu. However, since farmers only participated in the SLF with part of their arable land, the water saving per hectare of SLF households' total area is much lower than the above values.

3.2.2 Challenges of Implementing SLFP

Despite the progress made in SLFP, there are some problems challenging its effective implementation in the long term.

Some participating farmers were not qualified

Some participating farmers retired land themselves before the SLFP started. These farmers had already spontaneously fallowed land before being involved in the SLFP. Participating farmers should have had at least one plot of land on which winter wheat had been grown before, to be subsidized under the policy (Table 3.4). The table shows

Table 3.4 Farmers share of winter wheat area in total sown area the year before participation

	Share of winter wheat area in total sown area of a farmer interviewed (%)			Total
	0	Between 0 and 100	100	
Number of farmers	54	92	103	249
Share of farmers (%)	22	37	41	100

Source SLFP survey in 2019

that about one fifth of the farmers did not fulfill this requirement.

Fallowing land is economical for some farmers irrespective of a subsidy. The survey found that these farmers tended to stop growing winter wheat years ago, partly due to higher income from off-farm work. Another common reason is that they grow crops of higher economic value such as cotton instead of the wheat–maize succession. This means that some farmers, who should not have participated according to the project's policy, crowded out other farmers who would have really fallowed wheat—and thus saved water—through their participation.

Fallowing land was underused

The government encouraged farmers to grow green manure crops such as oilseed rape and alfalfa on land retired in winter and spring. However, only a fraction of the arable land involved in the SLF has been planted with such crops. Most fallowed land remains uncultivated, which may affect its fertility and result in a decline of production. In the NCP, the wind erodes the soil surface, especially in winter and spring when there is no plant cover and the winds are strong. Among a total of 2374 plots subsidized under the SLF project, 93.81% carried no crops in winter and 3.58% lay fallow for the full year. Only 2.61% of the plots carried some crops in winter, mainly oilseed rape.

Planting green manure crops helps to maintain and improve soil fertility while reducing water use. However, farmers rarely do so. There are several reasons. First, many farmers lack experience in planting green manure crops and publicity and scientific guidance provided by local government (on county or township level) are insufficient. Second, green manure crops generate costs for seeds, labor, and other items. It takes a long time for green manure crops to be converted into fertilizer in the soil, so in the short term the effect on improvement of soil fertility is slight or even not apparent at all. In addition, due to the late sowing time, seedlings of green manure crops may die through frost.

In conclusion, the willingness of farmers to plant green manure crops is very low, leaving the fallowed land underused. By forgoing the opportunity of improving soil fertility, the potential of SLF is not fully utilized.

Subsidy does not reflect the varying opportunity cost of land fallowing among farmers

The subsidy for fallowing is always 500 CNY/mu/year (7,500 CNY/ha), irrespective of the local circumstances, which means that the subsidy may be lower than farmers' expectation in some places and higher in others. Some studies show that the opportunity cost of SLF varies with the yield or the price of wheat. The yield of wheat in turn is affected by many factors such as soil quality and the availability of irrigation water. Therefore, compensation for fallowing in different areas should be adjusted according to the local conditions for agricultural production, in order to achieve fairness and efficient incentives for farmers in any area to be involved in SLF.

Policy sustainability is doubtful

Many farmers commented that if the policy ends, their retired land would be used to plant winter wheat again. In fact, the purpose of this policy is to encourage farmers to retire their winter wheat plots even if no subsidy can be provided in the future. The survey shows that 57.1% of households will plant winter wheat again after quitting participation in the SLFP.

When the subsidy decreases, the willingness of households to participate in the SLFP will decrease or even cease to exist (Table 3.5). When asked in the survey, the share of farmers willing to participate in the SLFP declined from 77% to less than 1% when the subsidy decreased from 500 CYN/mu/year to 100 CNY/mu/year. The issue is even more pronounced among participating farmers, 92.8% of which are content to take part in the policy under the current compensation standard. The willingness to participate decreased to 33.3% for a slightly lower subsidy of 400 CNY/mu/year, and none of the farmers was willing to participate if the subsidy declined to 100 CNY/mu/year (Table 3.5). So, farmers are very sensitive to the amount of subsidy. This implies that under the constraints of the government's budget for SLF, the sustainability of the policy needs to be carefully addressed.

After the SLFP, more than half of the participating farmers claim they no longer have the incentive to fallow land without subsidy, partly because their income is very low and fallowing land has still some—albeit small—impact on their income. When there is no more subsidy, they will plant winter wheat again to compensate for the loss of the subsidy. Take for example the allocation of time after fallowing land: in most cases, the time allocation was not affected by fallowing of land. Only 10% of farmers respond that the time they engaged in off-farm work contributing to their income increased (Table 3.6). This indicates that for most farmers, fallowing land has little impact on their life. They will adopt the previous planting mix and plant winter wheat when there is no subsidy.

In conclusion, the SLFP has without doubt brought about real water saving. It could be more efficient by adjusting amounts of subsidy to local conditions and by avoiding free riding. The most crucial point is its financial sustainability over time. At an avoidance cost of about 3 CNY per cubic meter of water saved it is a rather expensive measure (see also Box 5.1).

Table 3.5 Percentage of farmers willing to participate in SLF as a function of subsidy level (from CNY 100 to 500 per mu per year)

	Number of farmers	Subsidy (CNY/mu/year)				
		100	200	300	400	500
Non participant	311	1.6	6.1	8.7	15.1	64.3
Participant	249	0.0	6.4	13.3	33.3	92.8
Total	560	0.9	6.3	10.7	23.2	77.0

Source SLFP survey in 2019

Table 3.6 Influence on farmers' time allocation after participating in the SLFP

	Percentage (%)
Without influence	77.5
Leisure time increase	6
Time engaged in agriculture decreases	12.4
Time engaged in agriculture increases	0.8
Time engaged in off-farm work increases	10

Source SLFP survey in 2019

3.3   Farmers' Reaction to Policy Assessed Through a Groundwater Game

3.3.1   Introduction

During the Sino-Swiss project implementation, one item enthusiastically shared among the project team and the stakeholders was the groundwater game *Save the Water* (referred to as StW for short hereafter), which mimics the agricultural practice in the NCP. The development of StW resulted in two products, namely a board game version (Kocher et al. 2019) (Fig. 3.4) and a digital version (Fig. 3.5), respectively. The digital version is a web-based app, featuring real-time data transmission and the option to customize games. Detailed game instructions can be found in Appendix A-9.

Serious games originated in pedagogical fields (Apt 1970; Michael and Chen 2005) and have been increasingly used in public policy contexts, e.g. related to

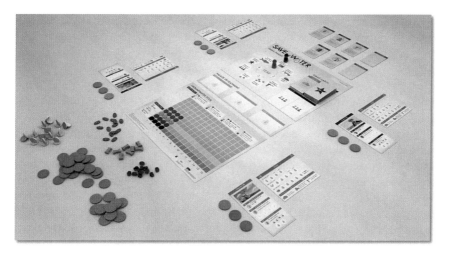

Fig. 3.4 The "Save the Water" StW board game

Fig. 3.5 User interface of the digital game "Save the Water" StW. The game can be accessed through the web link https://savethewater-game.com/game/

health care (Kato 2010), social morality (Katsarov et al. 2019) and, more recently, natural resources management, e.g. (Morley et al. 2017; Craven et al. 2017). Bots and van Daalen (2007) divide the possible functions of serious games for natural resource management in six categories, namely: (1) Research and analyse policy contexts as systems (game as a laboratory); (2) design and recommend alternative solutions to a policy problem (game as a design studio); (3) provide advice to a client on what strategy to follow in the policy process (game as a practice ring); (4) mediate between different stakeholders (game as a negotiation table); (5) democratize policy development by actively bridging stakeholder views (game as a consultative forum) and; 6) clarify values and arguments pertinent to the policy discourse (game as a parliament). Among those functions—given a sufficient correspondence between game and real natural system—using game as a laboratory allows researchers to draw valid conclusions from observations of the gaming process, which unfolds as players navigate through the game world. In addition, since the game world is a conceptual representation of reality, one can treat a player's decisions at different stages of a game as a reflection of his/her actual strategy under changing conditions of the real world. In this regard, a gaming episode is comparable with an interview, and hence the serious game may be used as a survey tool for data collection. Questions that one would ask in a questionnaire are now "answered" automatically and stored in a database as a game is played out. Compared with the questionnaire, the playfulness of serious games can make the "survey" process more enjoyable and, consequently, more motivating for farmers to participate in. Inspired by the literature, our team members carried out a field survey in Guantao using the digital StW game, and then conducted a behavioural analysis based on the game results to better understand Guantao farmers' decision-making and preferences in their agricultural activities.

3.3.2 Data and Materials

The game survey was conducted in October 2019, covering all eight townships in Guantao. For each township, two villages were selected from which 20 farmers were chosen. In total, 160 farmers were surveyed, including 26 farmers, who also participated in the CCAP's survey. Their profiles are shown in Fig. 3.6. The age of participants shows a bi-modal pattern, with one group concentrated around 40 years of age and the other around 55. It is consistent with the pattern from the Hebei survey. The majority of farmers received an education of nine years as mandated by China's national compulsory education program. The typical farm size is about 5 mu per family, and most participants have been farming for twenty to forty years. Despite the small sample size, the farmers' profiles are consistent with the previous SFLP survey by CCAP.

During the survey, each farmer first received oral instructions about the game rules, followed by a trial under customized easy mode to familiarize themselves with the StW game. At last, they proceeded to play formal games under the supervision

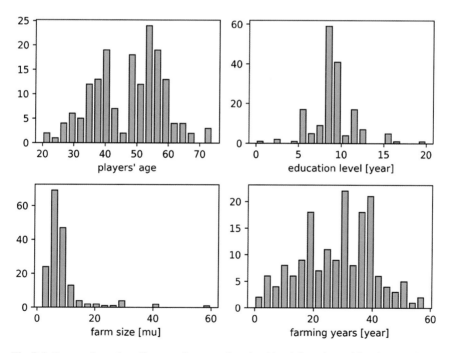

Fig. 3.6 Farmer players' profiles regarding age, educational level, farm size and farming experience

Fig. 3.7 Photos of farmer participants playing StW in the game-based survey

of students (Fig. 3.7). To compensate for their working hours lost, participants were granted a base subsidy plus an additional reward. This reward was set proportional to gaming results to keep participants motivated until they finish the games. Farmers were required to finish the game at least once. However, it was found that about 37% of players played more than three times.

The analysis of the gaming results has the goal to gain insights about players' underlying decision-making processes, especially the identification of factors that drive certain decisions.

In this work we adopt the decision-tree classification technique (Safavian and Landgrebe 1991; Breiman et al. 1984) for a behavioral analysis. Originating from the data mining field, the decision-tree classification fits a model on a sample by recursively partitioning the data into groups that involve instances of classes as uniform as possible. The structure of the model can be represented as a tree composed of nodes and branches, where the former corresponds to different features of a sample, and the latter are the splits of (sub-)samples. The derived rules are easy to interpret since they are merely "if…else" clauses that are, arguably, similar to the human decision-making process (e.g. Drakopoulos 1994; Drakopoulos and Karayiannis 2004). Therefore, the decision-tree classification can be used to formulate decision heuristics in modeling choice behavior, each rule stating a path of reaching a specific decision.

Table 3.7 List of main variables that characterize the StW game world and will be recorded in a database during a play. "Decision" refers to feasible actions that players can input in the game, while "Events" are external and random disturbances. They both affect the state of the game. Note that the weather conditions determine the amount of precipitation recharge to the aquifer

Category	Variable	Values
Decisions	Crop	Discrete choice among: "single crop", "double crops", "vegetables", "fallowing"[a]
	Farmland	Discrete choice between: "buy field" and "return"
	Amount of irrigation	A crop specific integer between 0 and 5
	Other input	Discrete choice among: "tractor", "water-saving irrigation", "greenhouse", "agricultural insurance"
States	Total capital	A positive integer
	Groundwater storage	An integer between 0 (empty storage) to 288 (the maximum storage)
	Current farm size	An integer between 1 to 36 (the maximum land size)
Events	Weather	"Dry year", "normal year", "wet year"
	Neighbouring water use	A positive integer depending on current water table
	Other	"Thunder strike", "pest damage", etc.

[a]Here "fallowing" means not cultivating any crop on a field. It is different from "seasonal fallowing" practiced in the NCP, where government subsidizes farmers for growing only summer maize instead of summer maize plus winter wheat in a year. In the game, seasonal fallowing corresponds to the choice of single crop

Since in StW most variables are of discrete type, it is straightforward to use decision-tree classification for the analysis, and the derived decision-tree may represent the discrete choice model of a player. Specifically, we use the Python implementation of the classification and regression tree algorithm (CART) (Pedregosa et al. 2011) as the classifier, which has been applied in behavioral studies of different contexts (e.g. Arentze et al. 2000; Su et al. 2017; Schilling et al. 2017; Huang and Hsueh 2010). The algorithm also provides measures for relevant feature importance as well as for the classification accuracy (Menze et al. 2009; Hossin and Sulaiman 2015). Table 3.7 shows a list of the main variables related to decisions, states, and random events in the game.

3.3.3 Results

Figure 3.8 summarizes the final performance of the games of all participants, with each line corresponding to one player. The performance is defined by a number of indicators (see Table 3.8). According to the plot, typical crop choices are single and double crops, with only few farmers growing vegetables. During the game survey, it was noticed that farmers unconsciously linked the game to their farming experience, and selected crop types based on what they actually grew in their farms. Moreover,

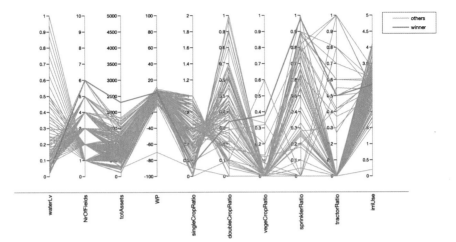

Fig. 3.8 Performance of players pooling results from all participants. Each line corresponds to the final performance of an individual player (yellow lines), with the best player who in the end owns the most assets highlighted in blue

Table 3.8 Definition of indicators used for comparing players' results. Among them, the "totAssets" indicator is used to define a winner within each policy group

Performance indicator	Definition
waterLV	Groundwater level in the last year of the game
NrOfFields	The total number of crop fields in the last year
totAssets	The total value of assets (capital and fixed assets) in the last year
WP	Average economic water productivity during a game
singleCropRatio	The ratio between single crop fields and total fields
doubleCropRatio	The ratio between double crops fields and total fields
vegeCropRatio	The ratio between greenhouse vegetable fields and total fields
sprinklerRatio	The ratio between fields with sprinklers and total fields
tractorRatio	The ratio between fields with tractors and total fields
irrUse	Average irrigation water use per field

only about 12% of players own a greenhouse. The irrigation behavior is also similar to reality, where Guantao farmers are used to irrigate twice for single crop and 3–4 times for double crops.

Regarding economic performance (i.e., totAssets and WP), results are clustered around the low end of the axis. In particular, the economic water productivity (WP) is around 10 and sometimes even negative, meaning that players did not exploit the value of groundwater to its maximum. In the StW game, acquiring more land plots is a key to the success of capital increase, and buying farming equipment such as sprinklers and tractors will further boost productivity. However, in the game farmers are rather conservative: they mostly get only one to two farm fields—similar to real-life households—and the adoption of tractors is low. In comparison, farmer players prefer to invest in sprinklers for saving irrigation cost, as shown by the high sprinkler ratio in the results.

Results of feature importance obtained from decision tree analysis are summarized in Fig. 3.9. For single crop decision (Fig. 3.9a), the crop chosen in the previous

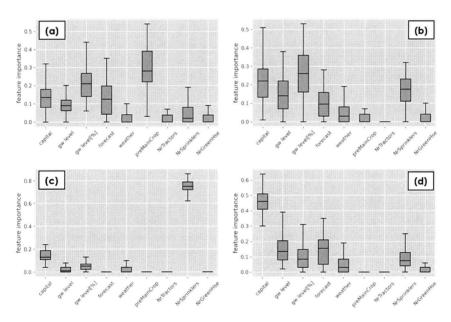

Fig. 3.9 Boxplots of feature importance for single crop decision (**a**), water saving irrigation (**b**), adoption of sprinklers (**c**) and land acquisition (**d**). the statistics of the box is computed from 50 times of training, each with a different sub-sample. Scores with larger value imply a higher importance in determing a specific decision. The notions of the factors along the x-axis are: "capital"—Money owned at decision-making step; "gw level"—groundwater level at decision-making step; "gw level[%]"—relative groundwater level at decision-making step; "forecast"—forecast weather at decision-making step; "weather"—actual weather at decision-making step; "preMainCrop"—the main crop type in previous step; "NrTractors" - the number of tractors possessed; "NrSprinklers"—the number of sprinklers possessed; "NrGreenHse"—the number of greenhouses possessed

decision time step (i.e., last "year") appears as a dominating factor. This implies that farmers' crop decision in this year strongly depends on what they did in the previous year, indicating a strong behavioral inertia. The capital level appears to have less influence on the decision. The second important factor is relative groundwater level, indicating that farmers do take into account groundwater availability in their decision-making process. Informing farmers about groundwater level, therefore, can be useful to promote single cropping (SLF in real life). Secondary to the factor of previous crop choice are capital level and weather forecast, which show a similar importance. This, on one hand, indicates that farmers' crop decision is not strongly economically motivated. On the other hand, farmers trust the weather forecast and use this information in planning their crop choices, even though the forecast accuracy in the game is uncertain.

Figure 3.9b shows the feature importance of "water saving" behavior. The "water saving" decision includes three situations in the game: (1) use deficit irrigation to save water at the cost of reduced income; (2) use sprinklers in the game to improve irrigation efficiency; and (3) do not irrigate at all. Results show that, in this case, the groundwater level [%] is the main factor that triggers players' water saving action. The capital level is the second important, possibly because each irrigation induces a cost. Sprinklers are the third important factor, which is expected since it saves irrigation water as defined by the rules: In StW, sprinklers can reduce a crop's water demand by one unit without affecting its yield.

Turning to the adoption of sprinklers (Fig. 3.9c), the number of sprinklers possessed has the highest importance, which is an intuitive result since it is less likely, even impossible, to buy new sprinklers during a game if one has already acquired many of them. But the results also confirm that the CART algorithm is working as expected. The second and third important factors are the capital and relative groundwater level, respectively. Therefore, if farmers have no sprinkler at all, the capital level will be the main obstacle that prevents them from adopting water saving equipment, while groundwater severity is the second concern. This might explain why programs such as subsidizing water saving equipment are welcomed by farmers, even though they have to pay maintenance cost themselves. Our results also suggest that, informing farmers about the severity of groundwater depletion could help to reinforce farmers' willingness in accepting programs for subsidizing water-saving equipment.

Regarding the land acquisition decision (Fig. 3.9d), the capital appears as the dominating factor, followed by weather forecast and groundwater level. Possibly because they are linked to groundwater availability for irrigation (In StW, buying new fields allows players to access more groundwater).

3.3.4   Discussion and Conclusions

The analysis of game results shows that farmers' crop choice has strong inertia and is less motivated by economic factors. The reasons behind such inertia are not clear, and can possibly be linked to the lack of experience with innovation, high age of farmers, risk averseness fearing failure when growing new crops or just to farmers' planting habits, since in NCP the single/double cropping has a well-established routine using mechanization. However, farmers in NCP do not buy but rather rent agricultural machinery from a company, which not only provides equipment but also service. Currently, smallholder farmers cannot afford to buy those machines themselves. One observation from the previous SLFP survey shows that farmers are willing to give up winter wheat for a subsidy of 500 CNY/mu, but few farmers take advantage of fallowed land to grow other permitted crops such as oilseed rape or alfalfa, even though such a practice can help to maintain the fertility of the soil. Therefore, unless changing crop structure is relatively cost-free (e.g., compensated by subsidy) and operationally easy, it will require a stronger economic incentive to persuade farmers to adopt new crops.

Water-saving irrigating behavior strongly depends on the groundwater level. Moreover, the groundwater level plays an important role in affecting other decisions in the game. Therefore, a proper communication with farmers about the severity of groundwater depletion can help to motivate water-saving behavior. In Guantao, the local water resource bureau has set up posters with historical records of groundwater head to show the declining trend. Also, the StW game can serve as such a communication tool for awareness raising.

The capital level is the main influencing factor for farmers' land acquisition and adoption of sprinklers—if one excludes the number of sprinklers possessed as a candidate determining factor. Therefore, if farmers have sufficient money, they are willing to invest on water-saving equipment. The higher capital level can also encourage farmers to buy new land, but from the results it is found that the land expansion behavior is rather conservative, with mostly only 1–2 new fields acquired during the game.

Using the game as a survey and analysis tool is a new concept proposed in our project, and the field experiments also suggests that farmers' decisions in the game show consistency with their real farming practice. Although the StW game omitted many nuances seen in real agricultural activities, such as the influence of seed quality, extreme weather other than precipitation, time length of irrigation, etc., it is nevertheless able to capture essential farmers' behavioral traits, and to provide directions for further investigation with formal econometric methods.

References

Apt CC (1970) Serious games: the art and science of games that simulate life in industry, government and education. Viking, New York

Arentze TA, Hofman F, Mourik H, Timmermans HJP, Wets G (2000) Using decision tree induction systems for modeling space-time behavior. Geogr Anal 32:330–350. https://doi.org/10.1111/j.1538-4632.2000.tb00431.x

Bots P, van Daalen E (2007) Functional design of games to support natural resource management policy development. Simul Gaming 38:512–532. https://doi.org/10.1177/1046878107300674

Breiman L, Friedman J, Stone CJ, Olshen RA (1984) Classification and regression trees. CRC Press

Cao G, Zheng C, Scanlon BR, Liu J, Li W (2013) Use of flow modeling to assess sustainability of groundwater resources in the North China Plain. Water Resour Res 49(1):159–175. https://doi.org/10.1029/2012WR011899

Craven J, Angarita H, Corzo Perez GA, Vasquez D (2017) Development and testing of a river basin management simulation game for integrated management of the Magdalena-Cauca River Basin. Environ Model Softw 90:78–88. https://doi.org/10.1016/j.envsoft.2017.01.002

Drakopoulos SA (1994) Hierarchical choice in economics. J Econ Surv 8:133–153. https://doi.org/10.1111/j.1467-6419.1994.tb00097.x

Drakopoulos SA, Karayiannis AD (2004) The historical development of hierarchical behavior in economic thought. J Hist Econ Thought 26:363–378. https://doi.org/10.1080/1042771042000263849

Feng W, Zhong M, Lemoine JM, Biancale R, Hsu HT, Xia J (2013) Evaluation of groundwater depletion in North China using the Gravity Recovery and Climate Experiment (GRACE) data and ground-based measurements. Water Resour Res 49:2110–2118. https://doi.org/10.1002/wrcr.20192

Feng W, Shum CK, Zhong M et al (2018) Groundwater storage changes in China from satellite gravity: an overview. Remote Sens 10(5):674. https://doi.org/10.3390/rs10050674

Hebei Government (2017). Notice of the People's Government of Hebei Province: announcement of Groundwater Overdraft Zones, Forbidden and Restricted Groundwater Exploitation Zones. Document Nr. 48 (In Chinese)

Hossin M, Sulaiman M (2015) A review on evaluation metrics for data classification evaluations. Int J Data Min Knowl Manag Process 5:1. https://doi.org/10.5121/ijdkp.2015.5201

Huang CF, Hsueh SL (2010) Customer behavior and decision making in the refurbishment industry—a data mining approach. J Civ Eng Manag 16:75–84. https://doi.org/10.3846/jcem.2010.07

Kato PM (2010) Video games in health care: closing the gap. Rev Gen Psychol 14:113–121. https://doi.org/10.1037/a0019441

Katsarov J, Christen M, Mauerhofer R, Schmocker D, Tanner C (2019) Training moral sensitivity through video games: a review of suitable game mechanisms. Games Cult 14:344–366. https://doi.org/10.1177/1555412017719344

Kocher M, Martin-Niedecken AL, Li Y, Kinzelbach W, Bauer R, Lunin L (2019) "Save the Water"—a China water management game project. In: Jonna Koivisto and Juho Hamari (Eds.) Proceedings of the 3rd International GamiFIN Conference, Levi, Finland, pp 265--276. http://ceur-ws.org/Vol-2359/paper23.pdf

Luo JM (2019) Evaluating water saving effects due to planting structure optimization in the Beijing-Tianjin-Hebei plain. Ph. D. thesis. Institute of Genetics and Developmental Biology, CAS, 2019

Luo JM, Shen YJ, Qi YQ et al (2018) Evaluating water conservation effects due to cropping system optimization on the Beijing-Tianjin-Hebei plain China. Agric Syst 159:32–41. https://doi.org/10.1016/j.agsy.2017.10.002

Luo JM, Qi YQ, Huo YW, Dong W, Shen YJ (2021). Optimizing planting structure for groundwater sustainability in Beijing-Tianjin-Hebei region, China: 1. Food supply/demand and water consumption. Submitted to Agric. Water Manag

Menze BH, Kelm BM, Masuch R, Himmelreich U, Bachert P, Petrich W, Hamprecht FA (2009) A comparison of random forest and its Gini importance with standard chemometric methods for the feature selection and classification of spectral data. BMC Bioinform 10:213. https://doi.org/10.1186/1471-2105-10-213

Michael DR, Chen SL (2005) Serious games: games that educate, train, and inform. Muska & Lipman/Premier-Trade

Mo XG, Liu SX, Lin ZH et al (2009) Regional crop yield, water consumption and water use efficiency and their responses to climate change in the North China Plain. Agr Ecosyst Environ 134(1):67–78. https://doi.org/10.1016/j.agee.2009.05.017

Morley MS, Khoury M, Savič DA (2017) Serious game approach to water distribution system design and rehabilitation problems. Procedia Eng 186:76–83. https://doi.org/10.1016/j.proeng.2017.03.213

Pedregosa F, Varoquaux G, Gramfort A, Michel V, Thirion B, Grisel O, Blondel M, Prettenhofer P, Weiss R, Dubourg V, et al (2011) Scikit-learn: machine learning in python. J Mach Learn Res 12:2825–2830. arXiv:1201.0490v4

Safavian S, Landgrebe D (1991) A survey of decision tree classifier methodology. IEEE Trans Syst, Man, Cybern 21:660–674. https://doi.org/10.1109/21.97458

Schilling C, Mortimer D, Dalziel K (2017) Using CART to Identify Thresholds and Hierarchies in the Determinants of Funding Decisions. Med Decis Making 37:173–182. https://doi.org/10.1177/0272989X16638846

Shen YJ, Zhang YC, Scanlon BR et al (2013) Energy/water budgets and productivity of the typical croplands irrigated with groundwater and surface water in the North China Plain. Agric for Meteorol 181(01):133–142. https://doi.org/10.1016/j.agrformet.2013.07.013

Su P, Yang J, Li Z, Liu Y (2017). Mining actionable behavioral rules based on decision tree classifier. In: 13th International conference on Semantics, Knowledge and Grids (SKG), IEEE, Beijing, China, pp 139–143. https://doi.org/10.1109/SKG.2017.00030

Wang J, Wang E, Yang XG et al (2012) Increased yield potential of wheat-maize cropping system in the North China Plain by climate change adaptation. Climatic Change 113(3):825–840. https://doi.org/10.1007/s10584-011-0385-1

Wang X, Li XB, Tan MH et al (2015) Remote sensing monitoring of changes in winter wheat area in North China Plain from 2001 to 2011. Trans Chin Soc Agric Eng 31(8):190–199

Xiao DP, Tao FL, Liu YJ et al (2013) Observed changes in winter wheat phenology in the North China Plain for 1981–2009. Int J Biometeorol 57(2):275–285. https://doi.org/10.1007/s00484-012-0552-8

Yuan ZJ, Shen YJ (2013) Estimation of agricultural water consumption from meteorological and yield data: a case study of Hebei North China. Plos ONE 8(3):e58685. https://doi.org/10.1371/journal.pone.0058685

Zhang XY, Chen SY, Sun HY et al (2011) Changes in evapotranspiration over irrigated winter wheat and maize in North China Plain over three decades. Agric Water Manag 98(6):1097–1104. https://doi.org/10.1016/j.agwat.2011.02.003

Zhang XL, Ren L, Kong XB (2016) Estimating spatiotemporal variability and sustainability of shallow groundwater in a well-irrigated plain of the Haihe River basin using SWAT model. J Hydrol 541:1221–1240. https://doi.org/10.1016/j.jhydrol.2016.08.030

Chapter 4
Decision Support for Local Water Authorities in Guantao

Policy selection and implementation rely on monitoring data and technical decision support tools. Monitoring data of Guantao County include groundwater levels at 55 observation wells, pumping rates of 7600 wells, surface water flows, precipitation, and land use in monthly time steps. The most difficult task is to monitor pumping of thousands of primitive irrigation wells. As all wells are powered by electricity in Guantao, monitoring via electricity consumption was chosen and shown to be effective. The electricity-to-water conversion factor is established via pumping tests. Four different groundwater models were developed serving different purposes: A box model computing the groundwater balance and the water gap of Guantao County; A distributed model of the shallow aquifer visualizing the spatial variation of groundwater levels and priority areas for pumping control; A real-time model updating and improving the distributed model by assimilating monthly observation data; And an upscalable data driven model using machine learning algorithms to forecast groundwater levels. All these tools together with monitoring data and current pumping control options are integrated in a web-based decision support system with a user-friendly interface. It allows local water managers without specialized knowledge to use complex technical tools in their groundwater management practice.

Electronic supplementary material The online version of this chapter (https://doi.org/10.1007/978-981-16-5843-3_4) contains supplementary material, which is available to authorized users.

W. Kinzelbach et al., *Groundwater Overexploitation in the North China Plain: A path to Sustainability*, Springer Water, https://doi.org/10.1007/978-981-16-5843-3_4

4.1 Hydrogeological Basis: Shallow Versus Deep Groundwater

Commissioned by the local project partners, a hydrogeological survey of Guantao has been carried out by China National Administration of Coal Geology (CNCG) in 2015. The contents of the report, which is directly relevant to our project is summarized in this section. It is augmented by some results from the hydrogeological mapping conducted by the China Geological Survey (CGS).

4.1.1 Borehole Locations

20 boreholes in total were drilled for this work to investigate the hydrogeological conditions in Guantao including 11 boreholes in the shallow aquifer and 9 boreholes in the deep aquifer (Fig. 4.1). The depths of the boreholes in the shallow aquifer vary from 60 to 100 m while the depths of the deep boreholes vary from 300 to 400 m.

Fig. 4.1 Borehole locations of the hydrogeological survey

4.1.2 Borehole Logs and Pumping Tests

Guantao is located in the Yellow River alluvial plain mainly composed of porous Quaternary deposits. In vertical direction, the deposits in Guantao can be divided into four aquifer layers interbedded with clay and silt aquitard layers. All aquifers are composed of permeable fine sand layers. The upper two aquifer layers are called "shallow aquifer", while the lower two are referred to as "deep aquifer" (Fig. 4.2). The average bottom level of the shallow aquifer is around 90 m below ground surface. The deep aquifer has a large thickness of around 200 m. The aquitard layer separating the shallow and deep aquifers has an average thickness of around 100 m. The piezometric heads in the shallow and deep aquifers differ by as much as 50 m. Furthermore, the hydrochemistry of groundwater is very different in both aquifers, with the deep aquifer being considerably less mineralized than the shallow aquifer. Hence, one can conclude that there is no direct hydraulic connection between the shallow and deep aquifers in the region of Guantao. The deep layer receives only little recharge from upstream at the piedmont, so groundwater levels have been decreasing continuously since pumping for domestic, industrial and irrigation uses started. The only way to stop the decline and start a recovery of the deep aquifer's piezometric levels is to abandon all pumping from the deep aquifer. The shallow aquifer is exclusively used for irrigation. It is recharged by precipitation, irrigation backflow and canal and river seepage. We focus mainly on modelling the shallow aquifer in the project.

Pumping tests were conducted at each borehole to investigate aquifer properties such as hydraulic conductivity (HK), and specific yield (SY) of the shallow aquifer. The HK-value of the shallow aquifer is around 10 m/day in the South of Guantao and decreases to around 1 m/day in the north. Similar results were found by the CGS in their mapping work. The SY values show a similar trend and decrease from South to North. Their absolute values could not be determined as the pumping tests were too short. Values from the literature were used.

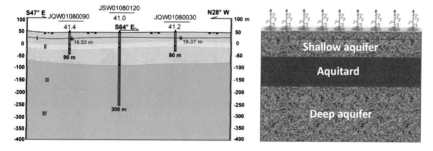

Fig. 4.2 Left: Hydrogeological cross-section of Guantao from west to east according to 3 borehole logs (Elevations in m asl. Brown: Aquitard, Green to Blues: Aquifers, I + II Shallow aquifer, III + IV deep aquifer) Right:Schematic view of aquifer structure generalized from borehole logs

4.1.3   The Distribution of Total Dissolved Solids (TDS)

The TDS distribution in the shallow aquifer is shown in Fig. 4.3 (left). The ground-water with TDS larger than 3 g/L cannot be used for irrigation. In these areas, groundwater from the deep aquifer is needed to irrigate crops. Therefore, the irrigation wells installed in the deep aquifer are mainly distributed in the areas with high salinity in the shallow aquifer (Fig. 4.3, right). The deep aquifer water is of drinking water quality with TDS-values clearly less than 1 g/L. Except for irrigation, the deep aquifer provides water supply for domestic and industrial use (Zone 2, the small area in purple color on the east boundary in Fig. 4.3 right). In recent years over 30 Mio. m^3/year of surface water were diverted into Guantao to replace groundwater used for irrigation, domestic and industrial purposes.

4.2   Monitoring Options

Monitored items include groundwater levels, pumping rates, land use and meteorological quantities. The monitoring data are transferred to a server, where they are used as input time series for model simulations and decision support.

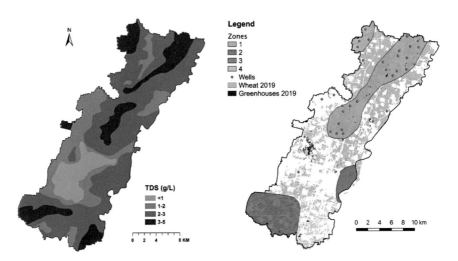

Fig. 4.3 TDS distribution in the shallow aquifer (left) and locations of surveyed pumping wells in the deep aquifer on the background of a satellite remote sensing map of wheat growing area and greenhouses (right). Wells in Zone 2 are urban drinking water wells. Zones 1, 3 and 4 have recently been connected to surface water canals

4.2.1 Groundwater Level Monitoring (Contributed by CIGEM and GIWP)

The groundwater level observation wells of the shallow and deep aquifers in Guantao are shown in (Fig. 4.4). The observation network was installed jointly by Hebei Department of Water Resources and the China Geological Survey, with most of the wells being recorded automatically on a daily basis. It comprises 55 observation points, 38 in the shallow aquifer and 17 in the deep aquifer. The modelling focus of the project is on the shallow aquifer (up to 100 m depth). Therefore, more observation wells are installed in the shallow aquifer. Since there is practically no recharge to the deep aquifer, head observations directly reflect pumping as the only item balanced by the long-term storage decrease. In 2017, it has been decided to close down all deep aquifer pumping wells. This has however only happened partially. The data obtained from the observation wells are discussed in detail below.

GW levels close to the river

The shallow observation wells ZK01-1, ZK02-1, ZK03, ZK04, ZK05, ZK06, ZK07, ZK08 and ZK09-1 of CIGEM (Fig. 4.5), lie on the eastern boundary of the county, on the two banks of the Weiyun River. The peak of observed heads in July/August 2016 is due to heavy rains and flooding, which means that the infiltration of water from the river is fast and leads to a groundwater level rise with practically no delay. After August, the "water mountain" under the river diffuses and the groundwater levels drop again. After 2016, groundwater levels fluctuate little as they are still mainly

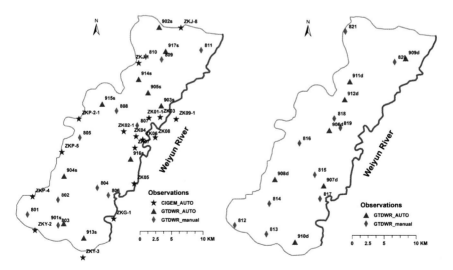

Fig. 4.4 Piezometers installed and monitored since 2015 within the project. Left shows the shallow aquifer observation wells, right shows the deep aquifer observation wells. Red stars: automatic wells monitored by CIGEM, blue triangles: automatic wells monitored by Hebei Province, Green diamonds: wells monitored manually by Hebei Province

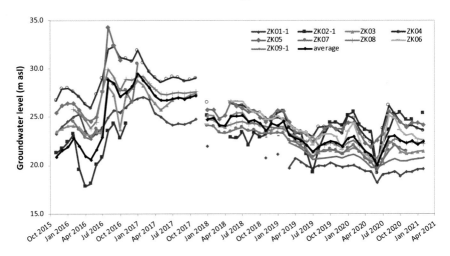

Fig. 4.5 Observed groundwater levels in the shallow aquifer recorded since 2015 by observation wells of CIGEM close to the eastern boundary of Guantao County

controlled by river infiltration. Since 2019, the groundwater levels again show large fluctuations as pumping for irrigation becomes the dominant driver, which is typical for normal and dry years. Due to little precipitation in 2019, the groundwater level shows a declining trend in that year.

GW levels on Guantao boundary away from the river

The heads recorded in the newly drilled observation wells of CIGEM in the shallow aquifer are shown in Fig. 4.6. These wells have been operating since 2017. The observed groundwater levels in wells ZKP-2–1, ZKJ-1, ZKP-5 and ZKY-2 are almost constant, which means that they are sufficiently far away from any individual pumping well and represent the outcome of the collective pumping in the surrounding area. The other wells show a decrease in groundwater level in the irrigation season and a recovery when pumping stops. The groundwater level in well ZKY-3 shows an increasing trend in 2020, probably caused by reduced groundwater pumping in that area. Generally, the groundwater levels on the boundary of Guantao represented by these observation wells have not shown any obvious decline since 2018.

GW levels inside Guantao boundary

The temporal pattern seen in the automatic observation wells of Hebei Province in the shallow aquifer (Fig. 4.7) shows the usual decline after the start of irrigation in March and recovery after September. In some locations, pumping activity in September/November is observed. Well 915s shows a constant groundwater level, comparable to what is seen in wells close to the western boundary in Fig. 4.6.

The deep observation wells of Hebei Province show a very similar pattern to the ones in the shallow aquifer (Fig. 4.8). Obviously, they supply irrigation water

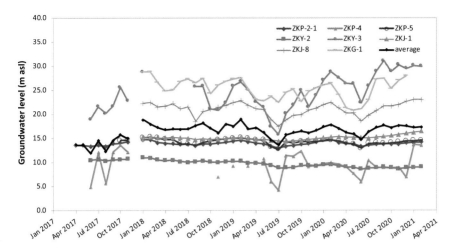

Fig. 4.6 Observed groundwater levels of the shallow aquifer recorded in new observation wells of CIGEM since 2017

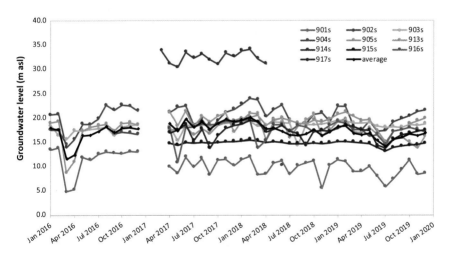

Fig. 4.7 Observed groundwater levels of the shallow aquifer recorded in automatic observation wells of Hebei Province since 2016

at the same time as wells in the shallow aquifer. The levels are generally lower and the amplitudes larger than in the shallow aquifer. As the deep aquifer is not recharged locally in Guantao, an increase in piezometric heads is either due to a decrease in pumping or due to pressure increase caused by the increased weight of the overburden in the course of large input of water at the surface (e.g. by extreme precipitation events as in 2016). Any decrease of a piezometric level is caused by pumping. The groundwater level decreased dramatically in 2019, due to intensified groundwater abstraction for irrigation in the drought year of 2019. Pumping in the

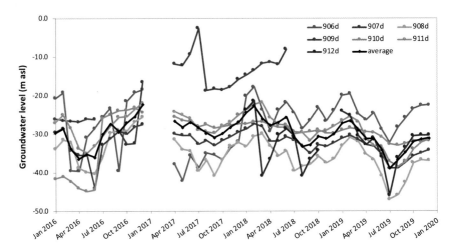

Fig. 4.8 Observed piezometric heads of the deep aquifer recorded in automatic observation wells of Hebei Province since 2016

deep aquifer has to be reduced to zero and the monitoring data can directly be used to verify, whether this goal has been reached.

The data of the observation wells of Hebei Province have been read manually, twice a year before 2015, three times a year in 2015, and four times a year after that. One can see that the levels of shallow boreholes gradually decreased from 2002, reached the lowest value in 2011, then started to recover and stayed practically constant between 2016 and 2018. After that they started to decline again (Fig. 4.9)

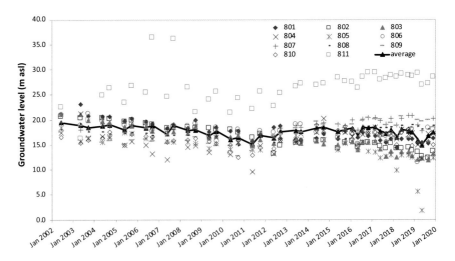

Fig. 4.9 Observed groundwater levels of the shallow aquifer recorded manually in observation wells of Hebei Province

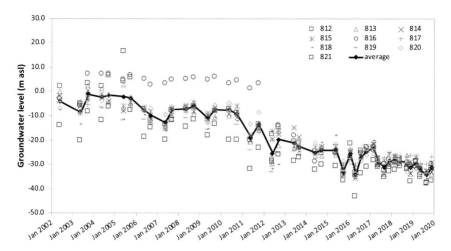

Fig. 4.10 Observed piezometric heads of the deep aquifer recorded manually in observation wells of Hebei Province

in the dry year of 2019.

The decline in observed heads of the deep aquifer is caused by both drinking water wells and irrigation wells. Before 2011 the levels showed a slow decline. From 2011 on, they declined considerably faster until 2017. Then the decline slowed down again (Fig. 4.10). The accelerated decline of some piezometers after 2011 is surprising. Possibly, there is an inhomogeneity in the data end of 2011. After 2017, the rate of decline in groundwater heads decreased due to the partial replacement of groundwater pumped for households and industry by surface water from the SNWT scheme.

Groundwater level contour map

In July 2020, a simultaneous measurement of groundwater levels of the shallow aquifer in Guantao and its surrounding areas was carried out by CIGEM to better characterize the regional groundwater flow direction. The groundwater level was determined in more than 200 wells (including some pumping wells). The groundwater level contour map of the shallow aquifer is shown in Fig. 4.11 (left), indicating that groundwater flows from east to west. The reasons are twofold: There is less over-pumping in the neighboring Shandong Province and the Weiyun River, which forms the eastern border of Guantao, supplies some recharge through river seepage. Two cones of depression are visible, one close to the western boundary in central Guantao, the other on the southern boundary, where the lowest groundwater level is located.

Similarly dense, simultaneous measurements of the groundwater levels of the deep aquifer are not available. The groundwater level contour map of the deep aquifer shown in Fig. 4.11 (right) is interpolated from the observed data of May, 2019 in the observations wells marked in Fig. 4.4 (right). It shows that groundwater in the deep

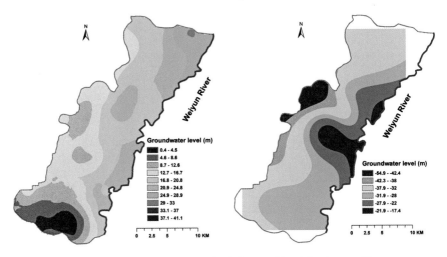

Fig. 4.11 Groundwater level contour map, left: shallow aquifer; right: deep aquifer

aquifer also flows from east to west. This can also be seen from the regional head distribution provided in Fig. 1.15. Piezometric heads are higher in the east.

It should be noted that the head distributions in Fig. 4.11 are momentary "snapshots". The general flow direction, however, does not change significantly over the year.

Conclusions

In recent years groundwater levels in the shallow aquifer stayed constant or decreased only slightly indicating reduced pumping compared to the time before 2014. This is due to two reasons: a large import of surface water after 2014 (Table 2.4), which amounted to about one third of the usual irrigation amount, and the fallowing measures implemented for winter wheat in Guantao since 2014 (Table 2.2).

The situation in the deep aquifer is different. While about half of the wells for drinking water and industrial water were decommissioned after their substitution with surface water from the SNWT scheme, agricultural pumping continued and even increased. It is essential for areas, where the salinity of the shallow aquifer is high and less-mineralized water from the deep aquifer is needed for dilution of shallow aquifer water or for sensitive greenhouse cultures. Consequently, groundwater heads in the deep aquifer layer are still declining. While the recharge-discharge gap in the shallow aquifer has practically been closed, the remaining gap is caused by the deep aquifer pumping. Paradoxically the increased pumping in the deep aquifer supports the groundwater levels in the shallow aquifer by irrigation backflow.

The equilibrium or even increase in heads in the shallow aquifer is very welcome. However, one should not become complacent as there are cycles of wet and drought years as seen in the past (Fig. 1.12). The next drought will certainly come, and it will be easier to manage if water levels in the shallow aquifer are allowed to increase now. Another important consideration is that with decreasing heads in the deep aquifer and with stagnant or increasing heads in the shallow aquifer the head difference between the two aquifers is further increasing. This implies an increased possibility of contamination of the high-quality water of the deep aquifer by leakage from the saline aquifer and aquitard above (Fig. 1.13). Finally, the continued recycling of irrigation water in a closed system such as a regional cone of depression in the shallow aquifer will inevitably lead to salt accumulation and thus deterioration of water quality. A sustainable system must provide an outlet for dissolved salts via drainage. Such outlets arise naturally if the drainage function of rivers and streams is revitalized by a sufficiently high water table.

4.2.2 Groundwater Pumping Monitoring

A water meter is the most straightforward method to monitor groundwater pumping. However, direct water metering at irrigation wells is still absent in most of the NCP despite huge investments by the local water authority. In comparison to the monitoring of groundwater levels, monitoring of abstraction is much more difficult, given the large number of wells. There are 7300 shallow pumping wells and 300 deep pumping wells. A low-cost metering solution would use cheap mechanical meters on each well, which are read off once a year. However, this solution is not deemed feasible by the water authorities. Previous experience shows that mechanical meters are unreliable and commonly tampered with. They often break under freezing conditions. They are not inseparably connected to pumps. They are removed when pumps are repaired, moved to another well, or completely taken out of the wells in winter.

A modern high-end solution is the smart water meter, which is offered for irrigation wells by several suppliers in China. A smart water meter is operated with a swipe card, which carries a prepaid amount of water. The volume pumped is abstracted from the amount stored on the card. The protection against tampering comes with the wireless connection of the smart meter to a server, receiving in real time all relevant data from the well usually at an interval of one hour. The transmission makes sure that the meter is working properly. If this is not the case, service personnel have to be sent to the field immediately to repair the meter. In China, such a system was first introduced in Minqin County, Gansu Province, in 2007 (Liu 2016). In the Sino-Swiss groundwater project's Heihe pilot site in Luotuocheng, Gansu Province, a smart metering system has been installed on 700 wells of the irrigation district in 2015, which is functioning very well up to today (Li et al. 2021b). The reasons for its functioning are fourfold: Wells have large capacities pumping around 150,000 m^3/year. They provide full irrigation under arid climate leading to a non-negligible water bill for the single user. The collection of a high water fee of 0.1 CNY for every cubic meter pumped provides sufficient funds for the maintenance of the system. Finally, smart meters have been classified as electrical equipment, making vandalism against them a crime punishable by existing law. None of these conditions is fulfilled in the NCP.

Equipment of more than 1000 wells in Guantao with smart water meters has been tried by Hebei Department of Water Resources with disappointing results. Of

Fig. 4.12 Typical irrigation wells and electricity meters installed in Guantao County

700 smart meters installed in 2016 only 6 were still working in 2018. While the initial installation was paid for by the water administration, funds for maintenance and repair were insufficient. Moreover, many wells are so primitive that they are not suited for the equipment with smart meters (Fig. 4.12). The main reason for failure was the farmers' lack of cooperation or even opposition resulting from unwillingness to pay water fees.

10 experimental smart water meters installed in this project provided the chance to assess the performances of different types of meters and to explore the feasibility of monitoring groundwater pumping by smart water meters on single wells. The measurements of the water meters allowed us to understand farmers' irrigation habits and the effectiveness of watersaving equipment.

Comparison of water meters

Five pairs of smart water meters from different producers were installed in Shoushansi District of Guantao. These meters differ in cost, meter type and data transmission method, which allows a comparison concerning the implementation of the different meters and measurement principles in practice. All the producers of water meters can provide the service of data management including data transmission and provision of a web-based data platform. A uniform communication protocol makes the data collection easier when several producers participate. An overview of the meter information is summarized in Table 4.1.

Mechanical water meters (Fig. 4.13) have the advantages of low cost, long service life and simple technology. When used for measuring groundwater, sand in the water can clog up and damage the propeller, causing a deterioration of measurement accuracy. Farmers generally resist the installation of sand filters as they entail an increase in energy consumption. Freezing can destroy meters immersed in water.

Table 4.1 Comparison of meters installed in Guantao for testing

Meter producer	Metering principle	Measurements	Price per unit[a]
Haisen (Tangshan, China)	Mechanical water meter (Nylon)	Water abstraction, electricity consumption, water level	4400 CNY
Hengyuan (Shijiazhuang, China)	Ultrasonic water meter	Water abstraction, electricity consumption, water level	12,000 CNY
Hengze (Qingdao, China)	Mechanical water meter	Water abstraction, electricity consumption	3000 CNY
RSA (Iran)	Electricity meter	Electricity consumption calibrated for flow	10,000 CNY
Itron (France/Suzhou)	Mechanical meter with high accuracy	Water abstraction	7600 CNY

[a]Not including cost of installation and modification of well necessary to house meter

Fig. 4.13 The Hengze meter (left), the Itron meter with transmission unit Watermind on the wall (middle) and the Haisen meter (right)

These are practical problems faced by mechanical water meters. The mechanical meters installed in the project differed in material, technology, and accuracy.

Meters based on ultrasound and electricity are shown in Fig. 4.14. Clamped-on ultrasonic water meters do not have the problem of clogging as mechanical water meters do, but the accuracy of the measurements can be impaired by air bubbles in the conveying pipe. Well houses or plastic huts are necessary accessories to protect the probes of the meters. The price of the meter and accessory equipment make ultrasonic water meters more expensive than mechanical water meters.

The electricity-based water meters only monitor electric energy consumption, which is then converted to water abstraction within the meter. An initial pumping test is required for its calibration. The test provides a conversion factor as a function of the pump's lift. Among the experimental monitoring systems, the electricity-based meter has the simplest hardware setup, containing only an electricity meter. However, the price of the meter from Iran is high due to the patented technology

Fig. 4.14 Hengyuan meter and RSA meter. Upper left: The Hengyuan ultrasonic water meter;Upper middle: The meter is protected by a concrete box. The control box and data transmission unit are protected by a plastic hut; Upper right: RSA meter box; Bottom: Swiss expert answering questions from the local farmers during the installation of RSA meters by technicians from RSA and local DWR

of data processing, which provides corrections of the conversion factor in time. A similar meter is available on the Chinese market for a considerably lower price from Huafeng Electronics Company, Jiuquan.

The experience of installing and operating these meters was very valuable, as it showed what could go wrong in the implementation of metering solutions regarding the issues of hardware installation, maintenance, data transmission and user-induced problems. Eight out of the ten smart water meters were not functional by the end of 2018 due to various reasons, listed below:

- Malfunctioning of the water meters.
- Farmers' unwillingness to cooperate and subsequent vandalism destroying the meters.
- Failures of data transmission units due to unforeseen update of the data transmission network or due to the malfunctioning of the transmission unit.
- Meters were bypassed manually due to circuit malfunction.
- Theft (One meter was stolen during the change of the well users.)

The meters that are still functional, are a clamped-on ultrasonic water meter and a high-resolution mechanical water meter. Although the operation of the ultrasonic water meter had been interrupted due to hardware malfunction and vandalism, its longer service time compared to other meters is attributed to the presence of an efficient local maintenance team. The high-resolution mechanical water meter is still functional at present. This type of meter is usually used by water utilities. It provides robust hardware and an independent battery-based data transmission unit, which makes it less vulnerable to vandalism. However, the cost of this type of meter is twice that of the common mechanical water meters.

The experience with the experimental water meters shows that a sustainable monitoring system must satisfy two requirements: (1) the hardware must be robust (for both meters and data transmission units) and (2) a prompt and efficient mainte-nance service must be provided. Guantao DWR had the same experience with their installation of smart meters on a larger scale.

Water usage and farmers' irrigation habits

From 2014 to 2016 in total 1037 water metering devices for pumping well monitoring have been installed in Guantao County through the locally funded projects. All these devices have been installed in wells with water saving equipment. In contrast, the 10 experimental water meters installed in this project have been installed in pairs on wells with and without water saving equipment (sprinkler) in the same region, to allow for an assessment of water saving, by comparison between the two, as well as farmers' irrigation habits. Annual water application per mu at five locations **with pipe irrigation** (flood irrigation in small field units) and five locations with **sprinkler irrigation** is shown in Fig. 4.15.

For each pair of measurements from the same meter brand, sprinkler irrigation utilized more water than pipe irrigation, except for the Hengyuan meter, which under-estimated the water use by sprinklers due to missing measurements. The average annual amount of water applied in sprinkler irrigation is $277 \, \text{m}^3/\text{mu}$, while the average annual water application with pipe irrigation is $263 \, \text{m}^3/\text{mu}$. The values are close to the water quota of $296 \, \text{m}^3/\text{mu}$, exceeding which farmers need to pay a water tax.

The surprising fact that sprinklers use more water than pipe irrigation is due to improper use of the sprinklers by the farmers. Flood irrigation in small field units requires application of at least 70 mm each irrigation time to flood the whole plot

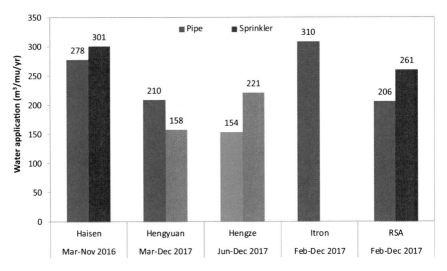

Fig. 4.15 Water application monitored by 5 experimental water meters (Bars in light colour indicate that meters covered less than one irrigation year)(Measurements of Itron sprinkler irrigation are missing due to malfunctioning of the pump.)(in m³/mu/year)

and water saving can only be achieved by reducing the frequency of irrigation. The water application by a sprinkler system is much more flexible. A sprinkler system in principle saves water due to its more even spreading. The application times can be adapted to plant need and water can be given more often in smaller amounts. As can be seen from Fig. 4.16, the irrigation days and the daily water application

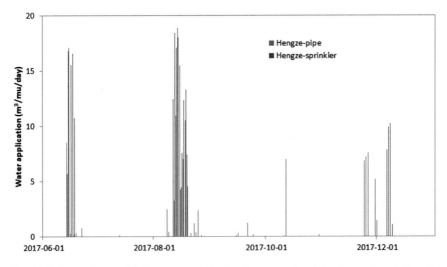

Fig. 4.16 Comparison of daily water application between pipe irrigation and sprinkler (in m³/mu/day)

of sprinklers are similar to those of pipe irrigation, which means that farmers apply sprinkler irrigation in the same way as they apply pipe irrigation, according to a fixed irrigation calendar. Therefore, no water was saved by using sprinklers. Efficient use of a sprinkler system means making full use of its flexibility in applying less water per time, spreading out irrigation times and targeting plant needs in critical crop growth stages instead of simply following the calendar days of flood irrigation. Possibly labor cost prevented a more efficient use.

It can also be observed that pumping lasts for one week (or less) in each irrigation season. Pumping observed in June and August was for irrigation of maize. Pumping in early October is usually expected to irrigate the winter crop fields before seeding. However, there was little pumping in October in 2017 due to sufficient precipitation in the previous months. Therefore, pumping is only observed by the end of November and in early December due to the dry winter.

Analysis without taking into account the particular situation in the field will produce biased results. Various factors can lead to a wrong determination of the amount of water pumped. For example, the Haisen meters did not work properly one year after installation. Still, the meter automatically transmits virtual data instead of real measurements when there is a hardware malfunctioning. For the user, this is difficult to detect as it can only be revealed when looking into the time series of flow rate in the data base. Additionally, when calculating water use per mu, errors can be introduced by inaccurate estimation of the irrigation area. This happens in two situations: (1) usually several households share one well, and the farmers or the electricians can only give a rough number of the total irrigation area of a well; (2) the irrigation area of one well varies as fields located between two wells A and B can be irrigated by either well A or well B. The error in the irrigation area is estimated to be around 20% according to farmers. The aforementioned factors prevent us from getting reliable results. These problems raise doubts in collecting water fees/taxes based on the monitored water use per mu.

Lessons learned

The smart metering experiment provided valuable experience in monitoring groundwater pumping at single wells. There is no evidence of obvious differences in the accuracy of the measurements using different types of water meters. The meters that monitor and convert electricity consumption to water abstraction have the advantage of being convenient and robust, but the accuracy depends on the initial calibration at each single well. All water meters require protection against tampering and theft.

The experiment raises questions about the feasibility of pumping monitoring based on smart water meters on single wells in well-intensive regions such as the NCP. First, the infrastructural investment of monitoring pumping on single wells will be enormous. Investment is not only needed for water meters and data management but also for upgrading and adapting the pipe systems at the primitive wells. Secondly, the monitoring system will not work functionally without efficient maintenance, which proved to be both costly and laborious. On-site inspection and data examination are two indispensable parts of maintenance work. The meter producers must offer to take on this task and cooperate closely with the local water department. Thirdly, farmers' unwillingness to cooperate had serious negative impacts on the operation of the monitoring system. Efforts are needed to educate farmers to accept the use of water meters and to properly use and protect them. The second and third requirements are hard to fulfil in the NCP, where there is little experience with public–private partnership in meter installation and maintenance yet, and where farmers are used to abstract groundwater without paying additional water fees.

The conditions are quite different from the situation in Europe, the US, Mexico and even Northwest China where agricultural wells are fewer in number, larger in capacity, and owned and controlled by a small fraction of the population. Monitoring and quantification of groundwater abstraction in these regions is feasible through direct monitoring of pumping activity. In contrast, any monitoring or control action on the small wells in the NCP involves millions of individual users, which greatly complicates an effective abstraction management.

4.2.3 Pumping Electricity Monitoring

A large number of pumping tests were performed from 2016 through 2018 to determine the conversion factor between electricity consumption and pumped volume. It was shown that for a single well with a single pumping test the annual pumped volume can be estimated from the annual electricity consumed within an error of less than 20% (Wang et al. 2020). The variation of the conversion factor from well to well is large with values varying between 1 m^3/kWh and 4 m^3/kWh for shallow wells. This implies that for equitability in fee collection a test has to be performed on each well. Technical background and procedures are provided in Appendix A-1.

To obtain an estimate of the total annual abstraction in a village at the same accuracy, it is sufficient to apply an average conversion factor, obtained from about 20 pumping tests distributed over the village, to the aggregated annual electricity consumption of all wells involved (Wang et al. 2020). In Guantao, the average conversion factors from electricity consumed to water pumped are 2.62 m^3/kWh for shallow wells, 1.32 m^3/kWh for deep wells and 13 m^3/kWh for pumping of surface water from the main irrigation canal to the fields.

Monitoring at single wells

All irrigation wells in Guantao County are equipped with electricity meters, which were mostly traditional mechanical electricity meters without data transmission before 2018, recorded once per month by the village electricians. Since 2018 Guantao EPSC has upgraded the metering system by installing smart electricity meters with data transmission at the frequency of once a day, which makes real-time pumping monitoring possible and allows to observe and identify the irrigation patterns of different crops (Fig. 4.17).

Due to the variability in the hydro-geological conditions, the types and working conditions of the pumps as well as the accuracy of the electricity meters, the spread in electricity-to-water conversion factors among single wells is large. Using a uniform

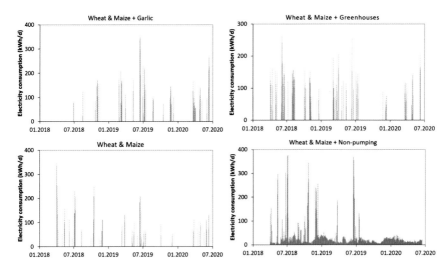

Fig. 4.17 Typical daily electricity consumption of single wells for irrigation of different crops in Guantao County (in kWh/day)

(average) conversion factor at single wells would lead to an error of up to 60%. Thus, for fee collection purposes, the conversion factor should be established for each well to achieve an accuracy of pumped volumes within 10–20%. Compared to using one uniform conversion factor, classifying the conversion factors by the pumps' rated powers improves the accuracy of water use estimation to a limited extent. It would be a compromise solution to increase fairness in collecting water fees. As pumping tests have not been conducted at every well, the average conversion factor calculated from pumping tests at representative wells has been used to obtain annual water volumes pumped at single wells. Not knowing accurately the irrigated area covered by a single well, it was suggested to collect water resources fees or taxes at village level. For this purpose, the use of an average conversion factor for the village would be justified. Using a uniform conversion factor on single wells is the equivalent of taxing energy use instead of water use, which is another method of taxing groundwater.

Quantification of water volumes pumped at single wells has been achieved by converting the annual electricity consumption of single wells shared by Guantao EPSC since 2017. Pumping electricity data include not only the electricity consumption of shallow wells but also energy consumption of deep wells and surface water pumping stations. By converting the electricity consumption using the corresponding electricity-to-water conversion factors, the spatial distributions of the use of different sources of water as well as the total use of irrigation water were obtained (Figs. 4.18 and 4.19). The use of deep groundwater is concentrated in Fangzhai, Chaibao, Luqiao and Weisengzhai Townships, where the salinity of the shallow groundwater is too high to be used for irrigation. The distribution of surface water consumption is concentrated along the Weixi Main Canal and branch canals in the southern and northern part of the county.

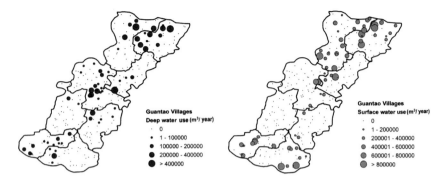

Fig. 4.18 Deep water use (left) and surface water use (right) of villages in Guantao in 2019 (in m³/year)

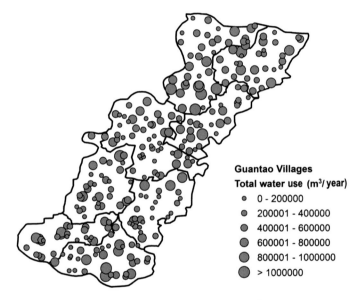

Fig. 4.19 Total water use of all villages in Guantao in 2019 (in m³/year)

The specific use of the different sources of water in the 8 townships of Guantao County is shown in Fig. 4.20. Shallow groundwater is the main water source for irrigation, while the fraction of deep groundwater is not negligible in most of the townships. On average, no township surpasses the limit of 296 m³/mu, while only 3 townships surpass the quota of 222 m³/mu. Although more than 30 Mio. m³ of surface water were imported to Guantao every year since 2014, the actual use of surface water only accounts for a small fraction of the total water use (similar to the situation in BTH described in Sect. 2.3). It is notable that the calculated total water use in some villages was greater than 1,000,000 m³/year, which is questionable and

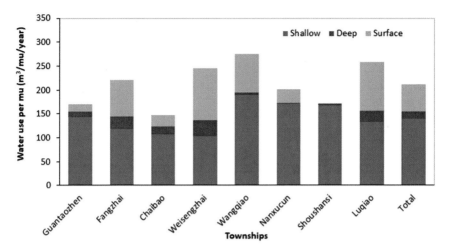

Fig. 4.20 Annual water use per mu of different townships in Guantao County in 2019 (in m^3/mu/year)

needs to be verified. This can be caused by incorrectly including some non-pumping electricity consumption by rural industry (e.g. chicken farms), connected to the same meter as the pumping well. Verification of the reported data is necessary to avoid overestimation of groundwater extraction.

Monitoring at transformers

At the final stage of the electric power distribution system, transformers step down the transmission voltage to the level required by the end users, which is 380 V (line voltage) for pumping wells. Power supply for agricultural use is separated from that of other sectors, i.e., residential and industrial uses, by using separate transformers. Electricity consumption is metered by Guantao EPSC at the transformers by smart electricity meters with remote data transmission once a day. There are 2 to 10 transformers for agricultural use per village. Each transformer supplies electric power for 3 to 10 irrigation wells. Monitoring at transformer level can certainly reduce the monitoring effort for a village by a factor of up to 10, from dozens of wells to a few transformers.

To test the monitoring method, 100 smart electricity meters were installed by the project on electrical transformers that supply power to irrigation pumps. The frequency of data transmission of these meters is once per 2.5 min, much higher than that of EPSC's meters. Since one transformer usually supplies power to several pumps, the meter on one transformer measures the total input power for the pumps connected to this transformer. The data collected, provide an opportunity to explore the possibility of monitoring several single wells' electricity consumption by one meter on their common transformer (Fig. 4.21). Machine learning techniques can be used to solve the data disaggregation problem involved. The method of Non-Intrusive Load Monitoring (NILM), which originated from household appliance monitoring

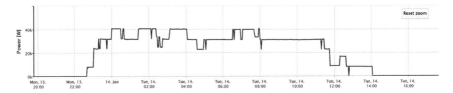

Fig. 4.21 Time series of the working power showing status of pumps connected to one transformer (In this example 4 pumps are connected to the transformer)

(Zoha et al. 2012), was applied to separate the electricity consumption of single wells from the total electricity consumption on transformers. Attempts were made to distinguish between types of pumps connected to one transformer, along with pumping duration. Analyzing events, where only one pump was operating at a time, led to successful separation of single pumps' contributions to the selected transformer's power output when no more than 4 pumps were involved. More sophisticated analysis is ongoing.

Monitoring at regional level

Historical electricity consumption records show that average pumping electricity for every month of the year is generally in line with the irrigation demand (Fig. 4.22). (Averages were taken over the time period from 2007 to 2016). One must note that the pumping electricity is not reported according to the calendar month, e.g., Shoushansi EPSA reads the electricity meter every 18th of a month. Thus, the peak of pumping electricity in April represents a peak of water consumption in late March and early April, which is the time of spring irrigation for winter wheat. Spring irrigation

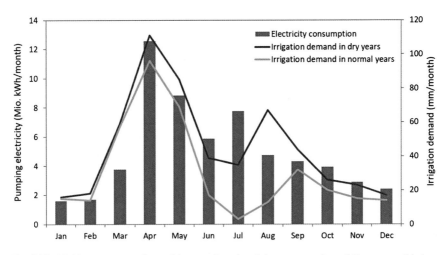

Fig. 4.22 Multi-year averaged monthly pumping electricity consumption of Guantao and irrigation demand for winter wheat-summer maize in dry and normal years (Averages of electricity consumption from 2007 to 2016 in Mio. kWh/month)

requires more water than the other irrigation seasons due to lack of precipitation in spring. It is interesting to observe that there is quite an amount of pumping in May and June, supposedly for irrigating winter wheat and vegetables. The third high value appearing in July represents the pumping for seeding summer maize in late June. There is no peak in October as one would expect during seeding of winter wheat. This can be explained by the abundant antecedent precipitation.

The inter-annual variability of pumping electricity of single districts agrees well with that of the county (Fig. 4.23). Only anomalies against multi-year average of pumping electricity are shown in the figure to avoid the influence of the magnitude of absolute values. The relative differences between the anomalies of single districts and that of the county are less than 20%, while the inter-annual variability of pumping electricity is much larger, with anomalies up to 50%, caused by the randomness of rainfall. It was also observed that the ratio between the pumping electricity of a district and that of the whole county is relatively stable from year to year. This implies that the pumping electricity of one region is also a good representative index of the pumping electricity, and consequently the irrigation water use, of a larger region. This indicates that it is feasible to estimate groundwater pumping of a larger area through monitoring of a smaller area.

As mentioned before, electricity-to-water pumping tests should be carried out for each individual well to reach an acceptable accuracy for taxing, while the required number of pumping tests can be much less, when estimating the groundwater abstraction of a larger area (village or county). Its accuracy depends on the accuracy of the areal conversion factor calculated as the average of tested conversion factors at individual wells. The results show that at a confidence level of 95%, at least 13 wells should be tested in Guantao if the error of the areal mean conversion factor of shallow wells is expected to be less than 20%. The number increases to 45 wells if the error is expected to be less than 10% (Wang et al. 2020). Compared to the total of about 7300 shallow wells in the whole county, the required number of pumping tests is very

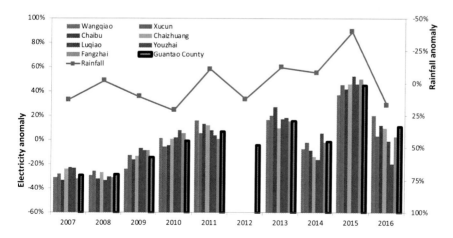

Fig. 4.23 Inter-annual variability of pumping electricity of the county and main districts

small. It implies that reliable areal estimates of aggregated groundwater abstraction can be obtained by testing only a small proportion of wells in the region. This conclusion provides the basis for reconstructing historical groundwater abstraction of the county for groundwater balance analysis (see details in the box model Sect. 4.3.1).

Recommendations

In Guantao, the depth to static groundwater table is large, on average more than 20 m. The cyclical fluctuation of the groundwater level within a year (on average ± 5 m) is small compared to the depth to groundwater table, which means that a pump's total lift does not vary much within a year, e.g., within 20% according to the measurements of conversion factors at a representative well. This situation is typical for over-pumped aquifers, where groundwater management is necessary. Under other circumstances, more tests may be needed to obtain a temporal distribution of the conversion factor over the year. The influence of a long-term groundwater decline is negligible within a year but becomes more significant over time. It has to be taken into account by repeating pumping tests every few years. Additional tests are also needed in extreme drought events when groundwater levels decline drastically beyond the normal range of pump lift or when a new pump is installed in a well.

Besides the errors introduced by energy-to-water conversion, there are also uncertainties in energy data collection. Although less critical compared to water meters, tampering is still a risk with traditional electro-mechanical meters. The introduction of new smart electricity meters with remote data transmission has eliminated this risk through efficient on-line detection. In China, power for irrigation wells is supplied by transformers for agricultural use, which are separate from those for industry and residential use. However, besides providing pumping energy, transformers for agricultural use also supply power for other agricultural activities, which are eligible for the subsidized electricity price of 0.5115 CNY/kWh, e.g., animal husbandry and preliminary processing of agricultural products. The regular price for households is tiered between 0.52 and 0.82 CNY/kWh. Additional efforts are needed to separate pumping electricity from non-pumping electricity in areas, where energy records of wells are not easily identifiable.

Cooperation between water and energy departments (DWR and EPSC) is crucial to achieve a successful system of groundwater abstraction monitoring using electric energy as proxy. EPSC has been sharing annual energy data at single wells with DWR since 2017 for water resources tax collection in all counties of Hebei Province. This is the first step of cooperation between electricity and water departments. Policy support and legislation are needed to achieve further multi-departmental cooperation to ease the procedures of data collection, data sharing, quantification of groundwater abstractions and implementation of control methods.

4.2.4 Land Use Monitoring

The planting area of the most irrigation intensive crops, winter wheat and greenhouse vegetables, can be monitored through satellite remote sensing. The relevant methods were developed or adapted in the project (Ragettli et al. 2018) (see also Appendix A-2). Figure 4.24 shows the winter wheat area of Guantao in the 5 consecutive years of 2016 to 2020 at a resolution of 10 m by 10 m. Some changes are visible, which are related to the policy of encouraging farmers through subsidies to fallow winter wheat. Winter wheat is easy to monitor as in January and February it is the only green crop in the fields. Except for haze days good images can be obtained in this season. The use of radar remote sensing products enhances the data base as it can "look through" the haze. The control of a cropping system by remote sensing alone has been practiced successfully in the Eastern La Mancha of Spain, where each irrigation unit assigns crops to each of its plots within its water right. The execution of the plan is verified by weekly satellite images (Sanz et al. 2016). The crop-specific temporal variation of the vegetation index week by week allows identification of the different crops and comparison to the crops declared. Only in cases of doubt an inspector has to do a check on the ground. Similarly, subsidized fallowing can be reliably verified in Guantao by remote sensing (Fig. 4.25).

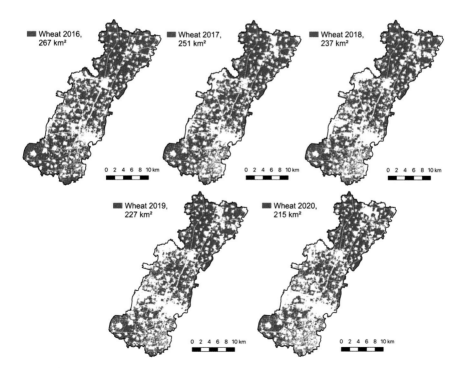

Fig. 4.24 Winter wheat area in Guantao County in five consecutive years 2016–2020

Fig. 4.25 Identification of fallowed wheat areas by comparison of images from two consecutive years

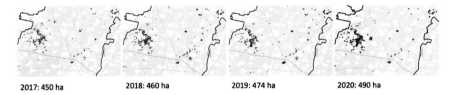

Fig. 4.26 Development of greenhouse area in Guantao (Image focuses on concentration in Shoushansi district) from 2017 to 2020

A relatively new development in Guantao is the fast increase in greenhouses, which are used for growing vegetables such as cucumbers and tomatoes. Although greenhouses save water for a single crop, greenhouse vegetables harvested multiple times per year are using more water per hectare than wheat–maize double cropping. Greenhouses can be detected by the spectral fingerprint of the plastic covers used (Yang et al. 2017). Figure 4.26 shows the development of the greenhouse area in Shoushansi irrigation district from 2017 to 2020.

4.3 Modeling for Decision Support

Four different groundwater models have been developed for interpreting the monitoring data gathered and to draw conclusions for groundwater management. Each of them has a different task. The first is a box model (0-D model), calculating a groundwater balance for the whole of the county (including shallow and deep groundwater).

This lumped model allows to determine the water gap for the whole county and the size of interventions necessary to close it. The second model is a 2D model of the shallow aquifer. It describes the spatial distribution of groundwater levels and allows to locate interventions (e.g. construction of infiltration basins, allocation of surface water supply) in an efficient way to fill up deep cones of depression. The third model is a real time version of the 2D model. It updates the 2D model continuously by assimilating monthly incoming observation data. From this model one can at any time get the best possible groundwater contour map of the county as a starting point for the calculation of scenarios. Not everywhere data are available in the same density as in Guantao. Therefore, the last model uses Guantao data to develop a machine learning procedure to predict groundwater levels at an observation well, which can be transferred to other locations. It is based on long-term records of past groundwater levels at the same well and all related time series data from its surroundings such as rainfall, irrigated area, pumping electricity use, etc. Guantao's deep aquifer is not modelled, as pumping has to be reduced to zero for sustainability and the monitoring data are sufficient to verify whether this goal is reached.

4.3.1 Box Model

A box model was used to analyze the water balance of the shallow aquifer in Guantao County, including abstractions from the deep aquifer. It can be used as a simple alternative to a spatially resolving numerical groundwater model to simulate the yearly change of the groundwater level averaged over the county. It is useful for fast scenario analysis in decision support. It can be transferred to other counties even if only standard data are available (see Box 4.1).

Model description

The box model considers the shallow aquifer of Guantao County as a whole and investigates the yearly inflows, outflows and change of storage. The scheme is illustrated in Fig. 4.27.

The recharge of the shallow aquifer includes infiltration of precipitation and irrigation backflow as well as seepage of surface water from the Weiyun River and the Weixi Canal system. Recharge components were estimated using empirical functions of the Handan Water Resources Assessment (Yang and Gu 2008). Pumping for irrigation is the only net outflow of the shallow aquifer. The lateral flux of the shallow aquifer is negligible compared to vertical fluxes (recharge, pumping), and so is exchange with the deep aquifer. Phreatic evaporation need not be taken into account, as the depth to groundwater has been larger than 10 m since the 1980s. This is well below the extinction depth for phreatic evaporation. It is also much lower than the bottom levels of river and canal beds. Therefore, surface water–groundwater interaction is a one-way-street: Surface water can seep in, but groundwater cannot drain to the river or irrigation canals. Pumping from the deep aquifer is accounted

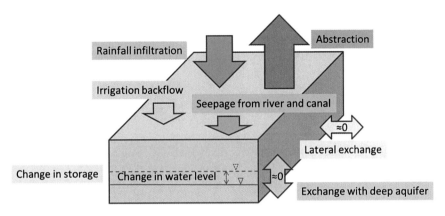

Fig. 4.27 Scheme of the water balance model of Guantao shallow aquifer

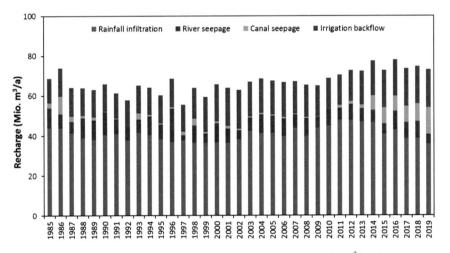

Fig. 4.28 Recharge components of the shallow aquifer in Guantao (in Mio. m³/year)

for, regarding both domestic and irrigation use. The backflow of irrigation with deep groundwater contributes to the recharge of the shallow aquifer.

Groundwater abstraction for irrigation as the only discharge of the shallow aquifer also influences the recharge to this aquifer by irrigation backflow. Data errors in groundwater abstraction greatly affect the results of the water balance calculation. Therefore, a considerable effort was made to quantify abstraction using pumping electricity as a proxy. Electricity records also allow the reconstruction of groundwater abstractions in the past. The electricity records that were used to reconstruct the total annual groundwater abstraction in Guantao County since 1984 include: 1) Single wells' electricity consumption from 2017–2019, 2) total electricity consumption for agriculture in Guantao from 2007 to 2016, and 3) rural electricity consumption in Guantao dating back to 1984, extracted from Guantao Statistical Year Books (SYB).

Power supply for agriculture is separated from that for other purposes, such as domestic and industrial use. Therefore, it should represent the electricity consumption for irrigation. However, considerable electricity consumption for other agricultural use, such as livestock and crop processing, was also included in the records of electricity consumption for agriculture. Before any conversion of electricity consumed to water volume pumped, it is important to exclude the non-pumping electricity consumption to avoid overestimation of groundwater abstraction. From total electricity consumption for agriculture, pumping electricity can be extracted by excluding the electricity consumption in non-irrigation months. From rural electricity consumption, pumping electricity can be derived under the assumption that the annual economy-related electricity consumption (for domestic and industrial use) increases smoothly with GDP while the inter-annual variability of rural electricity consumption reflects the inter-annual variability of pumping electricity due to the variation of precipitation from year to year.

Groundwater levels have declined drastically since the 1980s due to pumping for irrigation, which implies that the electric energy used to pump a unit of water has increased accordingly. Thus, the electricity-to-water conversion factor in the past has to take into account the smaller depth to groundwater. Water abstracted from both deep and shallow aquifers is used for irrigation, but the energy used to pump a unit of water from the deep aquifer is about twice the amount needed to pump a unit from the shallow aquifer. Thus, electricity consumption for pumping from the shallow aquifer has to be separated from that of the deep aquifer. However, before 2017 only the sum of both was recorded. To solve this problem, a parameter c, which denotes the ratio between the amounts of irrigation water abstracted from the deep aquifer and from the shallow aquifer, is introduced. It can be estimated from the irrigation areas of shallow wells and deep wells, under the assumption that the water use per unit area does not differ between fields supplied by shallow and deep wells respectively. More details concerning the box model can be found in Appendix A-3.

Water balance of Guantao shallow aquifer

The water balance of the shallow aquifer is determined by recharge and discharge. Rainfall infiltration is the main source of groundwater recharge followed by irrigation backflow (Fig. 4.28). The seepage from the canals was minor compared to other recharge components but has become comparable with the seepage from the river since 2014 due to the increase in imported surface water.

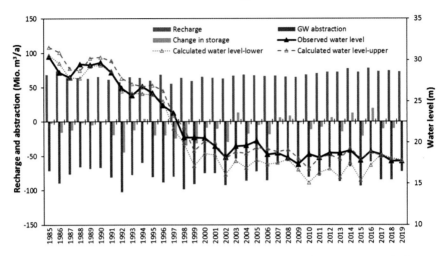

Fig. 4.29 Observed and calculated water levels and annual water balance components in Guantao County (in Mio. m³/year)

The calculated and observed water levels (averaged over the county) and the corresponding changes of storage in the shallow aquifer from 1985 to 2019 are shown in Fig. 4.29. Since only total pumping electricity was available before 2017, an uncertainty band of calculated water levels is presented to take into account the uncertainty of the parameter c. The range of c was set from 0.0 to 0.3. The annual variability of the calculated water levels generally agrees with the observations. The upper uncertainty band of the calculated water level ($c = 0.3$) basically coincides with the observations since around 2000, while the lower band ($c = 0$) is closer to the observations back in the 1980s and 1990s. This implies an increase in the fraction of deep groundwater use in recent years. It can also be partly attributed to an underestimation in converted pumping energy before the 1990s, when there were still some diesel pumps in use in irrigation. In the water balance calculations presented below c = 0 was used from 1985 to 1996, c = 0.3 was used from 1997 to 2016 and for the last years electricity use in irrigation was provided separately for shallow and deep aquifers.

Figure 4.29 indicates three periods of depletion of the shallow aquifer, a large water table decline rate between 1984 and 1999, a smaller one in the pre-project period from 2000 to 2013 and a still smaller one between 2014 and 2019. The corresponding depletion rates of the shallow aquifer are 15 Mio. m^3/year, 6 Mio. m^3/year and 1 Mio. m^3/year, respectively. The total gap between discharge and recharge including the deep aquifer is shown in more detail for the periods 2000–2013 and 2014–2019 in Fig. 4.30.

The components of recharge and discharge of the shallow aquifer for the two periods are shown in more detail in Fig. 4.31. In contrast to the relatively wet period between 2000 and 2013, three out of six years between 2014 and 2019 were dry years with annual rainfall less than 450 mm. This led to an increase in the abstraction from the shallow aquifer by 3%. However, the use of deep groundwater decreased by 35%, mainly due to replacement by imported surface water and fallowing. The increase in surface water use led to a drastic increase in canal seepage and partly contributed to the increase of irrigation backflow. Thus, the total recharge increased by 11% despite the decrease in rainfall infiltration.

The development of aquifer depletion in both shallow and deep aquifers for the two periods is shown in Fig. 4.32. The total gap could almost be halved during the project period (2014–2019) compared to the pre-project period (2000–2013). The shallow aquifer's recharge-discharge gap is down to 1.3 Mio. m^3/year from 6.4 Mio. m^3/year. The deep aquifer's gap is still 18.5 Mio. m^3/year but compared to the pre-project period it has been reduced by about 14 Mio. m^3/year due to project measures. If one takes the balance for the year 2019 only, the improvement is even more impressive. The total gap remaining in the deep aquifer is 11.2 Mio. m^3/year, while the shallow aquifer water level even rose, equivalent to a storage increase of 0.9 Mio. m^3. A single year is, however, not as significant for comparison as a period of several years due to the variability of meteorological conditions.

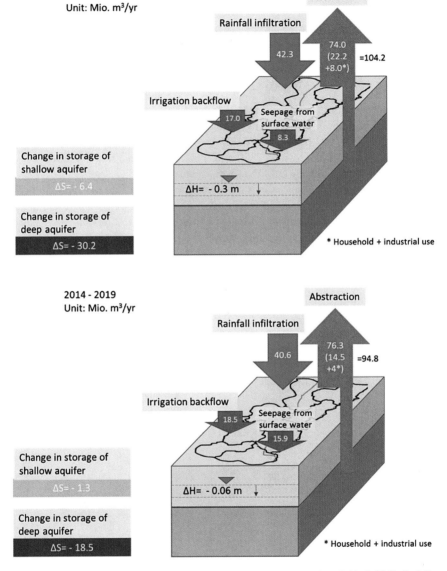

Fig. 4.30 Sketches of water balance of Guantao County 2000–2013 and 2014–2019 (in Mio. m^3/year)

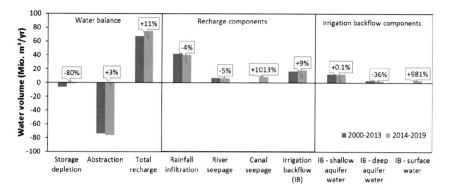

Fig. 4.31 Components of the shallow aquifer's water balance, 2000–2013 (dark blue) and 2014–2019 (light blue). Positive values for recharge (into the aquifer), negative values for discharge (out of the aquifer) (in Mio. m³/year)

Fig. 4.32 Guantao's aquifer depletion of different periods with contributions by irrigation and water supply for households and industry (in Mio. m³/year)

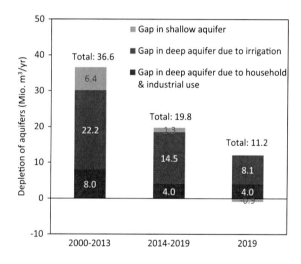

Transfer of the box model approach to other counties in the North China Plain

The box model can be used not only in Guantao but also in other areas with similar hydrogeological conditions. These conditions are:

- the net lateral flux of the aquifer is negligible compared to vertical fluxes (recharge, pumping).
- Exchange with the deep aquifer is negligible.
- Depth to shallow groundwater table is sufficiently large to exclude phreatic ET and exfiltration to surface water bodies.
- Groundwater is extracted by electric-motored pumps and is the main source of irrigation water.

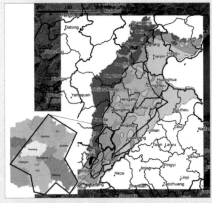

Basically, the whole of Heilonggang plain shown in the map fulfills these conditions. 8 counties of Handan prefecture served as a test (see insert in figure on the right). Data in Handan Reports of Water Resources were used as well as electricity data from Hebei and Handan statistical yearbooks.

Electricity data had to be corrected to remove non-irrigation use. Performance of the box model is acceptable in five out of eight counties Guantao, Guangping, Feixiang, Daming, and Cheng'an, where the inter-annual variability of water levels was captured most of the years. The poor performance of the box model in other counties is mainly attributable to the use of rural electricity instead of the actual pumping electricity. The reconstruction of pumping electricity for total rural electricity consumption is inaccurate due to the increasing fraction of non-pumping electricity consumption (for domestic and industrial use) in total rural electricity consumption. The performance of the box model will improve when pumping electricity is available for a more accurate estimation of groundwater abstraction, preferably separated into shallow and deep aquifer pumping.

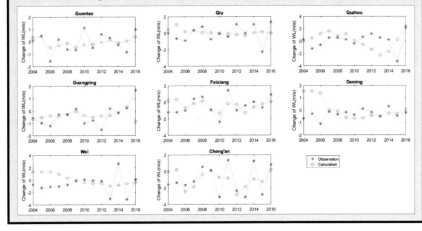

Box 4.1: Upscaling of box model approach to Heilonggang low plain region in the North China Plain

4.3.2 Distributed Groundwater Model for Guantao

A numerical groundwater flow model facilitates the understanding a local hydrogeological system. It serves as a basis for decision making by allowing to evaluate and compare outcomes of different water allocation scenarios in a spatially resolved manner. As explained in the previous chapter, we focus mainly on the shallow aquifer. Therefore, a 2D groundwater model of the shallow aquifer in Guantao is developed using a structured grid discretization.

Model development

The groundwater model of Guantao is based on USGS Modflow (Harbaugh et al. 2000) under the GUI Processing Modflow (PMWIN) (Chiang and Kinzelbach 2000). The county area is discretized into 210 rows and 190 columns with a grid size of 200 m × 200 m. The top and bottom of the aquifer are determined by the digital elevation map and interpolated borehole logs, respectively. The model parameters including hydraulic conductivity, specific yield and recharge ratio are determined by data from hydrogeological maps, soil maps, and pumping tests. The model boundary condition is a specified head boundary determined by interpolation of the observed groundwater levels in the observation wells on the boundary. The main sources of groundwater recharge in Guantao are precipitation and irrigation. The recharge is assumed to be a constant fraction of the sum of the two fluxes. Groundwater pumping is the main sink term in the numerical model. It is calculated by converting the pumping electricity consumption of each village by a conversion factor (see Sect. 4.2.2). The Weiyun River on the east boundary infiltrates groundwater as well. Its infiltration is not described explicitly in the numerical model, but implicitly contained in the specified head boundary condition along the river, which is influenced by the river's infiltration.

The numerical model is running in transient mode for the period between March 2018 and December 2019 in monthly time steps. The initial condition of the transient model is interpolated from the measurement data recorded in February 2018, using the Kriging method. After transferring source/sink terms to each model grid, uncertain model parameters are adjusted to minimize the residuals between the observed and computed head values. More details about the model set-up can be found in Appendix A-4.

Model results

The calibrated hydraulic conductivities are shown in Fig. 4.33. The value decreases in south-north direction. There is a relatively impermeable zone between HK1 and HK3 with a value of 3.3 m/day. Like the hydraulic conductivity, the calibrated specific yield decreases from south to north (Fig. 4.34). The specific yield influences both the amplitudes of head time series over the year and the general trend over the longer term. The calibrated recharge ratios are shown in Fig. 4.35. The values tend to decrease from east to west, which is consistent with the local hydrological observations.

The correlation coefficients among different calibrated parameters are presented in Table 4.2. Correlation coefficients with a modulus above 0.6 indicate that there is a non-negligible correlation among the respective parameters. The highest correlation coefficient between two zonal parameters is 0.6 (between HK1 and SY1). This implies non-uniqueness of the two parameter values. An increased HK1 with a

Fig. 4.33 Distribution of
hydraulic conductivity (in
m/day)

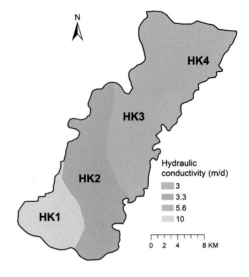

Fig. 4.34 Distribution of
specific yield

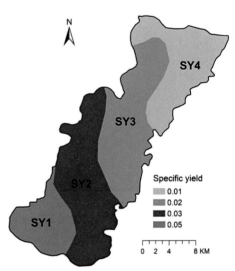

simultaneously increased SY1 will lead to a similar model result. The smaller correlation coefficients show that all other hydraulic conductivities and specific yields
are relatively independent, and their relative values can be uniquely determined by
calibration.

The groundwater balance of the transient model is presented in Fig. 4.36. The
total groundwater recharge is about 0.22 Mio. m^3/day, which contains an average
recharge from irrigation backflow and precipitation of around 0.16 Mio. m^3/day
(73% of the total groundwater recharge). The infiltration from precipitation and
irrigation fluctuates gently due to the delay and damping effect of the unsaturated

Fig. 4.35 Distribution of recharge ratio

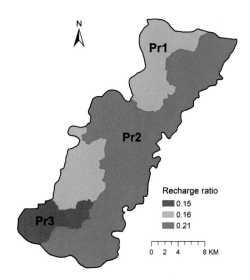

Recharge ratio
- 0.15
- 0.16
- 0.21

0 2 4   8 KM

Table 4.2 The correlation coefficients among different calibrated parameters

	HK1	HK2	HK3	HK4	SY1	SY2	SY3	SY4
HK1	1.00	0.03	0.00	0.00	0.60	−0.04	0.00	0.00
HK2		1.00	0.03	0.01	0.03	−0.15	−0.02	0.02
HK3			1.00	0.00	0.00	−0.02	0.01	0.16
HK4				1.00	0.00	−0.01	−0.06	−0.39
SY1					1.00	−0.11	0.00	0.00
SY2						1.00	−0.02	0.00
SY3							1.00	−0.39
SY4								1.00

zone. The boundary inflow and the main canals' infiltration are not shown in the figure because they account for a small part of the total groundwater recharge (only 11% and 16% respectively). The pumping rate fluctuates strongly. It follows the temporal behavior of precipitation, which directly (and without delay) influences the amount of groundwater pumping used for irrigation. The average pumping rate is about 0.2 Mio. m^3/day.

The aquifer storage increases when groundwater recharge exceeds groundwater withdrawals and is depleted when the groundwater withdrawals exceed groundwater recharge. The groundwater storage starts to deplete in March and recovers after August. The average storage depletion is about 0.025 Mio. m^3/day or 9 Mio. m^3/year. The depletion in 2019 is large as 2019 was a very dry year. Over-pumping is defined as the longer-term average depletion, which is clearly smaller. To a certain degree groundwater levels inside Guantao are influenced by the larger surroundings via the boundary heads. This influence from the outside can be separated from the influence

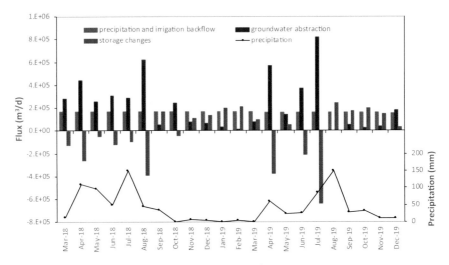

Fig. 4.36 Water budget of 2D groundwater model (in m³/day)

by pumping in Guantao itself. Li et al. (2019) show that on average inside and outside contribute about equally to groundwater level changes.

The comparison of modelled and observed groundwater heads over time is shown in Fig. 4.37. Some piezometers underestimate and others overestimate observations. The seasonal pattern of groundwater level dynamics, however, is reproduced in most observation wells. The groundwater levels start to decrease from March due to irrigation and reach their lowest values in July/August. From July/August on, groundwater levels start to recover and increase as pumping ceases and recharge from precipitation and irrigation backflow dominate. Some observation wells (wells 901s, 907sd, o801) close to the boundary fail to capture the dynamics of groundwater levels because they are controlled by the specified values of observations on the boundary, which did not show any strong dynamics. As recharge from precipitation and irrigation backflow varies gently in time, the intra-annual groundwater level dynamics are mainly caused by pumping of groundwater from the shallow aquifer. Since the shape of the irrigation area of each village is not known exactly, it is defined by constructing Thiessen polygons around the village centers. It is quite possible that in this process, an observation well falls into the wrong village in the model, and thus captures different groundwater level dynamics. Besides, some observation wells are not strictly used for observation only. They are pumping wells which are monitored only once or twice per month when the pump is switched off. Therefore, it is inevitable that data monitored just before or right after an irrigation event show larger dynamics than they should.

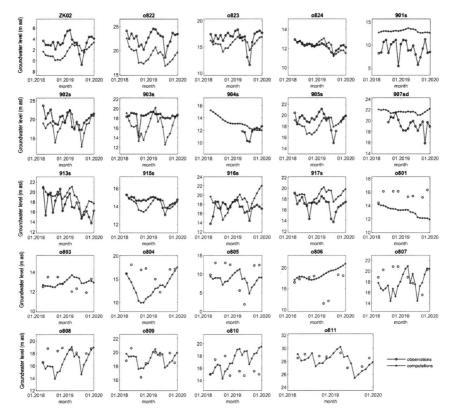

Fig. 4.37 Comparison of calculated and observed monthly time series of groundwater heads at different observation wells

4.3.3 Real-Time Groundwater Model

Real-time groundwater modelling, also known as data assimilation, is a dynamic model updating procedure incorporating all data monitored in real time such as groundwater levels, surface water diversions, groundwater pumping, etc. as they become available. It uses an ensemble technique (Ensemble Kalman Filter or EnKF) to update model parameters and states considering the uncertainties in parameters/states/inputs. The model also allows to make a forecast, which is continuously improved on the basis of new observations.

Uncertainty quantification

The model parameters updated in the real time model include the 4 zonal hydraulic conductivities, 4 zonal specific yields and 3 zonal recharge infiltration ratios shown previously. The parameters are assumed to follow log-normal distributions with a standard deviation of 0.5. The measurement error is assumed to follow a standard normal distribution with a standard deviation of 1 m. The pumping rate is assumed to have a normal error distribution with a standard deviation of 20% of the mean value.

Procedure

The principle of the updating process includes: (1) generating fifty replicates for each model parameter/state/input from the assumed distributions; (2) running the numerical model 50 times with different combinations of parameters and inputs; (3) updating model parameters and states by assimilating observations available at this time step using EnKF; (4) inserting the updated parameters/inputs into the numerical model for a model forward run in the next time step; (5) repeating steps 2–4 until the end of the assimilation period. The procedure is applied to the 2D model of the shallow aquifer of Guantao. There are 22 monthly assimilation time steps. A damping factor (Hendricks Franssen and Kinzelbach 2008) is introduced in the EnKF to correct spurious model updates and forecasts of the error covariance. Without damping, the procedure converges before having collected enough experience regarding uncertainty and cannot follow larger head changes (Li et al. 2021a). More details concerning the real-time model are provided in Appendix A-5.

Results

Figures 4.38 and 4.39 represent the results from applying the EnKF method to update parameters and states. The uncertainty of parameters and states decreases with time due to the assimilation of observation data. The dynamics of the ensembles follow the observed groundwater level change very well. However, the ensemble in the northern well cannot completely capture the real observation values during the winter months. This indicates that there is still some bias between the numerical outputs and real observations, which cannot be explained by the presently assumed model uncertainties. The spatial distributions of groundwater level means and standard deviations are shown in Fig. 4.40. Groundwater depression cones have formed in Guantao, which have a dominant influence on the groundwater flow direction. The

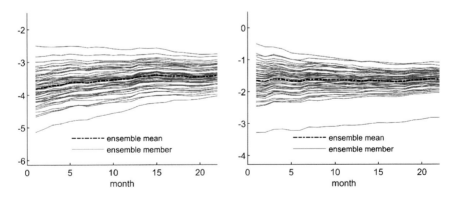

Fig. 4.38 Evolution of typical parameters' probability distribution (log value) (Left: log(SY3), right: log(Pr2))

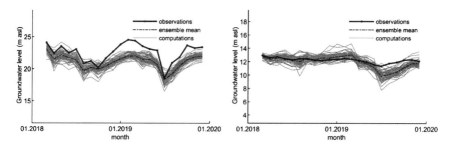

Fig. 4.39 Typical ensemble trajectory (Left: observation well in the north; right: observation well in the south)

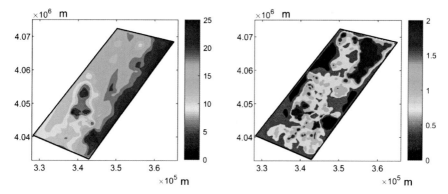

Fig. 4.40 Groundwater contour maps of July, 2019, for ensemble mean (left) and standard deviation (right)

areas with relatively larger standard deviation indicate larger uncertainty. This is due to fewer or no observations in these areas (see observation location map Fig. 4.4 in Sect. 4.2.1). It is obvious that new observation wells have to be installed to reduce uncertainty, especially in the Northeast corner of Guantao.

Conclusions

The model results show that the assimilation of data in a model has two advantages over traditional modeling: First, it allows a sequential improvement of model parameters with incoming data. Second, it gives results for states and parameters together with a quantitative estimate of their error. The estimate of the error of groundwater heads shows how good the model is in predicting states in the following time step. The real time model has been developed and tested with past observations. In the application, the prediction for the next month is of major interest. The prediction interval can be increased, however at the cost of accuracy. In a predictive mode, the model can be used in operational decision-making, e.g. management decisions before determining winter crop planting in autumn. Irrespective of all parameter uncertainty and prediction uncertainty, the real time model with data assimilation presents the best possible use of all information. The map of the ensemble average heads after assimilation is the best head contour map one can obtain. This map is imported into the decision support system as basis for planning. It serves as initial conditions for scenario calculations and in its form of map of distance to groundwater (which is obtained by subtracting the head map from the digital terrain map) one can see where corrections in pumping in Guantao County are needed most urgently.

4.3.4 Data Driven Groundwater Model

Traditionally groundwater level forecasts are computed with either box models or numerical models. They require various types of data concerning model structure, model parameters and other model inputs as explained before. Recently, with the development of automatic monitoring, head data sets of higher frequency and greater length are available in real time. Therefore, instead of using complex, physically based models with a large number of model inputs and model parameters, more and more researchers explore the use of data driven models in predicting groundwater levels using a more limited number of data items but with much longer time series of each. Besides the predictions, a quantification of their uncertainty is provided. For more details refer to Appendix A-6.

Data

Data driven modelling builds a functional relationship between given pairs of input and target output datasets (i.e., samples). Based on the derived relationship, forecasts can then be made for new input data. In this study, the Support Vector Regression (SVR) (Basak et al. 2007) method is used for the prediction of groundwater levels in Guantao County. The data used in Guantao for groundwater level forecasts include monthly groundwater level data, monthly rainfall data, monthly surface water diversion data and monthly agricultural electricity consumption data without correction for non-pumping electricity.

Model development

We develop data driven models for each of 3 observation wells and their average, which represents the whole of Guantao. The time series data sets are divided into three subsets: training data set from 2007 to 2015, validation data set in 2016 and test data set from 2017 to 2018. The input variables are wavelet decomposed into sub series (Fig. 4.41). The time lags regarding the relevant input variables are preliminarily determined according to a hydrological analysis. The groundwater levels were time lagged by 12 months and the time lag of other input variables varied from 1 to 12 months. Thus, there are 37 input variables in total before wavelet-decomposition.

In the study we use both non-wavelet decomposed input variables and their wavelet decomposed components as potential model inputs to select the best features (Quilty et al. 2019). During input feature selection, only the top features, providing more

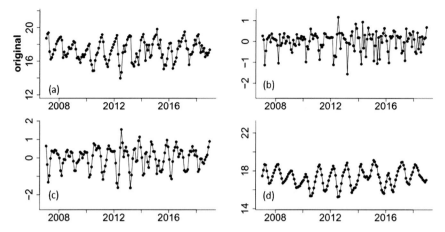

Fig. 4.41 Original groundwater level time series and the corresponding decomposed sub-time series ((**a**) is the original time serieis, (**b**) is the high-frequency sub-time series at the first decomposition level; (**c**) is the high-frequency sub-time series at the second decomposition level; (**d**) is the low frequency sub-time series at the second decomposition level)

information to the forecasts than the rest of the features, are retained for the development of the models. We use 100 bootstrap resamples to quantify the uncertainty in the input variable selection and in the model parameters. The groundwater levels are predicted for lead times of 1 month, 3 months and 6 months, respectively.

Results

- Groundwater level prediction of the whole of Guantao

After training SVR models, the groundwater levels are forecasted for 1, 3 and 6 months ahead using the test data. The quality of forecasts at the different lead times is quantified for different SVR models for the three lead times 1, 3 and 6 months. Note that the optimized numbers of input features for the three lead times are 4, 5 and 5 respectively. The groundwater level predictions for the average groundwater level over the whole county are provided in Fig. 4.42.

The root mean square error (RMSE) and the Nash–Sutcliffe efficiency coefficient (NSE) are applied to quantify the model error. The results show that the ensemble mean fits observations well even though some forecasts over- or underestimate the observations. The 1-month lead time predictions have the lowest prediction error and provide the best fit. It is not surprising that the model performance degrades for predictions with larger lead times. Going from lead time of 1 month to lead time of 6 months, the RMSE of the groundwater head predictions averaged over Guantao County increases from 0.6 m to 0.78 m. The NSE decreases from 0.6 to 0.32.

- Groundwater level predictions for each single well

Table 4.3 shows the model evaluation metrics of groundwater forecasts for the average of the 3 wells (representing the whole of Guantao) and for each observation

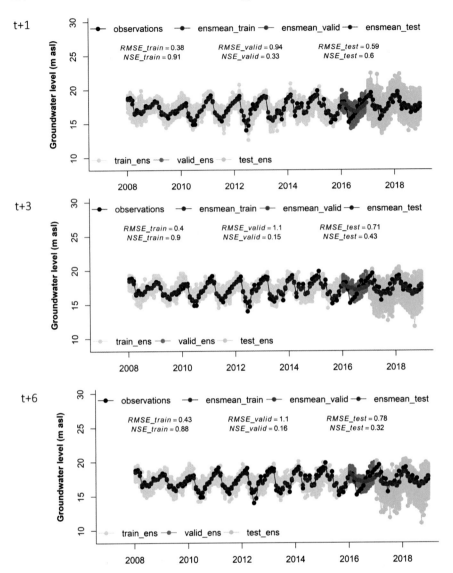

Fig. 4.42 Groundwater level prediction at different lead times (t + 1,t + 3 and t + 6). The gray, black and green shadows show the ensembles of groundwater levels calculated for the training period, the validation period and the test period, respectively

well separately (well N in the north, well M in the middle and well S in the south of the county) at 1, 3 and 6 months lead times, respectively. The stochastic groundwater level predictions for each single observation well lead to similar conclusions. As in the case of the county average, the 1-month lead time predictions at each single well yield a lower prediction error and better fit than the predictions for larger lead

Table 4.3 Model performance for groundwater level predictions at lead times of 1, 3, and 6 months in the test period

Locations	Index	t + 1	t + 3	t + 6
Whole of Guantao	RMSE (m)	0.59	0.71	0.78
	NSE	0.60	0.43	0.32
well N	RMSE (m)	2.14	3.48	3.75
	NSE	−0.95	−1.75	−4.99
well M	RMSE (m)	0.74	0.97	1.09
	NSE	0.63	0.37	0.2
well S	RMSE (m)	0.62	0.83	0.89
	NSE	0.07	−0.76	−1.04

times. The prediction for each single observation well underperforms compared to the results for the whole of Guantao (having higher RMSE and lower NSE).

The prediction error decreases from north to south at all lead times. The RMSE decreases from 2.14 m in well N to 0.62 m in well S at 1 month lead time, from 3.48 m to 0.83 m at 3 months lead time and from 3.75 m to 0.89 m for 6 months lead time. The model at well M has the highest NSE of all three wells, implying that the forecast's mean fits observations quite well at well M. The lower accuracy of model predictions for each single well may arise from the fact that the groundwater level in the single well may not be representative for the local area where the other, area-averaged inputs are collected (e.g. precipitation). Besides, the groundwater level in the single well may also be influenced by lateral fluxes over county boundaries, which are not considered as inputs here.

Discussions and conclusions

A data driven model using a machine learning algorithm is a general way to forecast groundwater levels. Practically, there are also other conventional alternatives to forecast groundwater levels, such as long-term average (S1), long term average plus the average dynamics in a year (S2), simply using the data of the previous month as predictor (S3), and the ensemble mean from the data driven model (S4). Here we take the forecasts at 1 month lead time as an example. The results are shown in Fig. 4.43. The long-term average scenario cannot reproduce annual dynamics (or patterns). S2 considers the annual patterns averaged over the past, and therefore can make a better forecast when the real pattern fits the historical dynamics (e.g. in 2017 but not in 2018). S3 produces better results than S1 and S2 due to the inertia of groundwater level change. But S4 still performs slightly better than S3. The scenario S4 produces the smallest root mean square error (RMSE) of groundwater levels with a value of 0.6 m. The RMSEs for S1, S2 and S3 are 1 m, 0.88 m, and 0.8 m respectively.

In conclusion, the data driven model using a machine learning algorithm is recommended in practical applications. Future improvements include the use of longer time series and the use of agricultural electricity consumption data corrected for non-pumping use. This section demonstrates that one can easily implement data driven

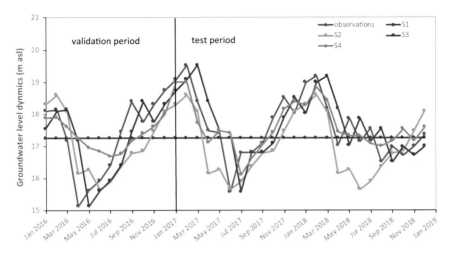

Fig. 4.43 Groundwater level predictions of different scenarios

models to forecast groundwater levels. The method can be applied to any area in the NCP, provided sufficiently long data sets are available. It should however be noted that a model trained with data from the past cannot predict behavior due to events, which never happened within the available time series such as an extreme rainfall, the sudden allocation of large amounts of surface water, or a severe change of the pumping regime due to policy change.

4.4   An Integrated Decision Support System (DSS)

4.4.1   Motivation

During the project implementation, a number of technical products were developed, including various groundwater models, a crop water demand model, algorithms for land use identification and, most importantly, a database for storing all information collected so far. Those tools have shown to be very useful in the analysis of the over-pumping issue in Guantao, and in exploring various mitigation strategies. However, each tool was developed as a separate software resulting from each team member's own research task. But soon it became clear that the lack of coupling between the modules did not correspond to their interconnectedness in real life and thus limited the efficiency of their use. Therefore, in the second phase of the project, one main task was the seamless integration of those tools into a unified framework to assist groundwater management in Guantao. This naturally led to the idea of implementing a decision support system (DSS).

Decision support systems (DSSs) refer to computer aided programs that help decision makers to solve unstructured or semi-structured problems using data and models (Morton 1971). They are widely used in different fields nowadays. Among them, environmental decision support systems (EDSS) specialize in tackling environmental problems and have attracted a growing attention from researchers, e.g. (Matthies et al. 2005; Hadded et al. 2013; McDonald et al. 2019; Whateley et al. 2015; Shao et al. 2017). The popularity of EDSS, on one hand, is due to the increasing difficulties in solving environmental problems, which are often entangled with human society, turning them into complex coupled human-nature systems (Liu et al. 2007). The complexity is further amplified by global change which imposes extra uncertainties concerning the future state of the world (Pachauri et al. 2014; Walker et al. 2013; Milly et al. 2008). On the other hand, the improvement of computer and monitoring technologies have enabled researchers to produce/access vast amounts of information and to develop more advanced analytic tools. However, those trends not only increase the potential of EDSS, but also demand modern EDSS implementation to be more democratic, user-friendly, and flexible (Mir and Quadri 2009; Hewitt and Macleod 2017; Loucks 1995; Zulkafli et al. 2017). The democratisation implies the participation of stakeholders in data collection, development of EDSS as well as their final operation. The user-friendliness aims at lowering of technical barriers so that policy makers lacking specific knowledge can still themselves apply those analytic tools for their decision making. The flexibility requires EDSS to be able to include new information/functionalities with ease but also to allow integration with legacy models, which were built with state-of-the-art science but not designed for the Internet context (Kumar et al. 2015). Web-based applications have become a popular solution that overcomes many of the challenges such as accessibility: Users can easily visit applications on their computers or smart devices (Swain et al. 2015). The benefit can be further leveraged via cloud computing to remove computation limits, and to provide on-demand load balancing in the face of high numbers of simultaneous users.

In this project, we implement such a web based DSS, named Guantao Decision Support System (GTDSS), using modern information and communication technology (ICT). The GTDSS aims to achieve two goals in particular, as shown below:

(1) **Monitoring**: Informing decision makers about states of the system based on observation data and model estimates, where the states include groundwater level, cropping area, water use and hydro-climatic conditions, etc.

(2) **Planning**: Designing management strategies according to sustainability criteria and the target performance specified by decision makers.

The target users are officers from Guantao local water bureau, which is a small department. In addition, they have limited experience in using professional modelling tools, and poor access to ICT technical support—usually outsourced to 3rd party companies. To reduce the technical burden in using our system, the GTDSS is co-designed with stakeholders, implementing a user-friendly operability by adopting interactive visualization and adapting the model configuration to their needs. In addition, the GTDSS is based on open-source software and is launched as a web site hosted on a Chinese cloud provider, namely the AliCloud. Users can access the system via the links: https://www.ifu-gwm.com or http://www.giwp-gtdss.cn. The webpage serves as a graphical user-interface (i.e. front-end), which can be opened with any modern browser (Safari browser is currently not supported). On the web site, users can perform various operations, such as running models or query monitoring data. All operations initiated by end-users are handled by the server (i.e. back-end), on which the models and the database are installed. The tools and development workflow can also be applied to other cases for implementing environmental decision support systems (Li 2020).

4.4.2 Key components

GTDSS is composed of five key modules:

- Water quota planning
- Scenario analysis
- Irrigation Calculator
- Crop mapping, and
- Data portal

Each module is implemented as individual web page, which can be accessed from the navigation panel on the top (Fig. 4.44). In addition to that, the **Home** page

Fig. 4.44 Home page of Guantao Decision Support System (GTDSS)

provides brief descriptions of the project as well as the functionalities of each module, while the **Others** page introduces miscellaneous resources such as the groundwater game, the project's and partners' websites, etc. The following sections will give a quick glance at each module and explain its duty. Readers can refer to the Technical Guidebook in Appendix A-8 for more information regarding the instructions of use and technical details.

Quota planning

This module allows users to plan shallow groundwater pumping and winter wheat fallowing for the next year according to a *target depth to groundwater table* and *expected surface water supply*, as input by users. The *target depth* can only be set within a feasible range defined by two "red lines", namely an upper and a lower bound, respectively. The upper bound mandates the groundwater level to be no higher than the extinction depth (5 m below ground level) to prevent soil salinization from phreatic evaporation. The lower threshold accounts for a minimal level to ensure that the reduced pump yield from a lowered groundwater level is not less than 70% of the originally designed pump yield, to which a reserve groundwater storage is added, sufficient for mitigating irrigation water shortage over a design drought of 4 years duration. In Guantao, these requirements lead to a lower red line at 30 m below ground surface. The user interface (UI) is shown in Fig. 4.45.

As output, the module informs the user about the amount of reduction in pumping volume eventually required. This implies a comparison of the calculated sustainable pumping volume with last year's pumping volume. If less shallow groundwater can be used, the quota planning module further suggests the area of winter wheat fallowing needed in order to achieve the required water saving.

The suitable time for quota planning is September, because: (1) the winter wheat planting hasn't started yet; (2) 90% of annual precipitation has fallen; and (3) the

Fig. 4.45 The interface of Quota Planning module. Panels 1–3 inform users about key information on groundwater management drawn from observations. Specifically, panel 1 shows the estimated spatial distribution of depth to groundwater table, panel 2 the observed recent average depth to groundwater table over Guantao and the county's water use, and panel 3 measurements of other key variables. Panel 4 implements input options for quota planning and the output window for displaying results

surface water supply is known. Under these conditions the uncertainty of model inputs is reduced, and decision makers have sufficient time to prepare the budget for the seasonal fallowing program.

Scenario analysis

This module implements two groundwater models, namely the box model (or 0-D groundwater model) and the distributed numerical model (or 2-D groundwater model), which were described in Sects. 4.3.1 and 4.3.2.

The inputs for the 0-D groundwater model are the expected surface water supply and the expected pumping electricity use. The output is the predicted (county) average depth to groundwater for the next year. The box model UI is shown in Fig. 4.46.

The 2-D groundwater module allows more comprehensive input options, as shown in Fig. 4.47. For example, users can specify driving forces such as precipitation and surface water supply, "build" ponds for artificial aquifer recharge, and, most importantly, specify inputs for water allocation. In addition, users can define crop water demand and irrigation area for different crops, based on which the model automatically estimates total water use. To take advantage of the spatially distributed nature of the model, the total water use, land use settings and surface to groundwater irrigation ratio can be defined at township level. The output of this module is the prediction of head development over the next 10 years under the given scenario, assuming that conditions remain the same on average over the simulation horizon.

Fig. 4.46 UI of box modeling tool. Panel 1 is the input panel to define simulation scenarios, and panel 2 is the output of depth to groundwater table time-series, with uncertainty bounds for the forecast given in red

Fig. 4.47 Interface of the 2-D groundwater modelling tool. Panel 1 provides comprehensive input options to specify boundary conditions and the allocation settings on a township basis. For each township, users can also define landuse and crop water demand via panels 1a and 1b, respectively. Panel 2 is the output window for the spatial map of simulated groundwater level change. Panel 3 displays the time-series of level change at a selected location, as well as the water allocation results per township

Irrigation calculator

This module implements the core routine of the Irrigation Calculator App, which is an on-line software that simulates crop water demand under different climate and land conditions. Its capabilities are shown in Box 4.2 and Box 4.3 and described in detail in Appendix A-7. It can be accessed via the links https://app.hydrosolutions.ch/ IrrigationCalc-Guantao2/ or http://www.giwp-gtdss.cn/. As shown in the top panel of Fig. 4.48, the main inputs to the Irrigation Calculator are soil type, crop type and planting date, while the output is crop water demand, either at a monthly time step or aggregated into the annual total. The same model is also used in the 2-D groundwater model introduced in the previous section for providing reference crop water demand.

Fig. 4.48 Interface of the Irrigation Calculator module. The top panels are the UIs for estimating crop water demand based on the crop model, where panel 1 is the input window and panel 2 shows outputs. The bottom panels are the UIs for looking up official values defined in the irrigation norm issued by Hebei government. Users' query parameters are specified via panel 3, and the query results are shown in panel 4

Guantao Irrigation Calculator

A tool for monthly irrigation water demand in Guantao County, Hebei, China

An important task of water managers around the world is to deliver water for irrigation when it is needed. Traditionally this is done by consulting irrigation norms that provide annual estimates of irrigation water demand depending on crop and soil types and on irrigation method and farm size. However, the distribution of the irrigation water throughout the growth season of the plant is often not included in the irrigation norms which makes the planning for timely water deliveries to the farms difficult. Within the frame of this project, a web application was developed, which allows the monthly planning of irrigation water demand: Guantao Irrigation Calculator.

Procedure

- Choose local average climate or define user specified climate
- Choose local crop types and enter planting dates
- Choose among local predominant soil types
- Choose typical local conveyances and application losses

Output

- Monthly irrigation demand
- Comparison with local irrigation norms
- Saving & loading of irrigation planning tables
- Available in Chinese & English

Monitoring of crop activity

- Archive of weekly enhanced vegetation index (EVI)
- Real-time computation of average EVI
- Real-time computation of time series of average EVI
- Comparison to time series of regional average EVI

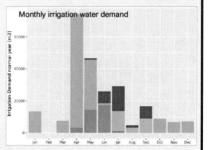

Reference: Allen, R. G., et al (1998). Crop Evapotranspiration. FAO Irrigation & Drainage Paper http://www.fao.org/3/X0490E/x0490e00.htm

Web access to Guantao Irrigation Calculator: https://app.hydrosolutions.ch/IrrigationCalc-Guantao2/

User manual for the Guantao Irrigation Calculator: https://www.hydrosolutions.ch/projects/guanto-irrigation-calculator

Video tutorial: https://youtu.be/rCeiuTboorY?list=PLrox336x2AMQy5qBfyPPLZmT1rVM82oHB

Box 4.2: The Guantao Irrigation Calculator (contributed by Beatrice Marti)

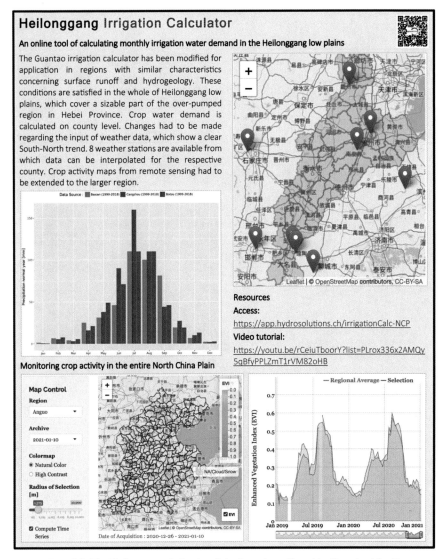

Box 4.3: The Heilonggang Irrigation Calculator (contributed by Beatrice Marti)

In addition, the module also includes an option for looking up the irrigation norm of a crop from the database, which digitizes the official document (DB13/T 1161.1-2009). It specifies the reference water quota for main crops in different regions of Hebei Province, and gives adjustment factors taking into account the type of water source, the size of irrigation area and the irrigation method. The interface is shown in the bottom panel of Fig. 4.48.

Crop mapper

The original CropMapper is an online application that allows users to generate high resolution crop maps by running a state-of-the-art machine learning algorithm on the Google Earth Engine platform (https://hydrosolutions.users.earthengine.app/view/cropmapper-ncp). For details see Appendix A-2. It utilizes publicly available remote sensing images as inputs. This module periodically retrieves the irrigation map outputs from the full-fledged CropMapper app and stores them into the database of GTDSS, which can then be used for monitoring the irrigation area for a specified crop type and year (see Fig. 4.49). The crop types include summer crop, winter crop, maize, winter wheat and greenhouse vegetables.

Data portal

This module allows users to download/visualize various data related to Guantao groundwater management, as shown in Fig. 4.50. Table 4.4 lists the variables included in this module and accessible features. All data are extracted from the database.

Fig. 4.49 Interface of the CropMapper module. Panel 1 provides input options to query different crop maps for different years. Panel 2 is the output window where the extracted crop map will be rendered on the base layer of Guantao region. Panel 3 displays the time-series of crop areas at township level

Fig. 4.50 Interface of the data portal module. Panel 1 shows the data inventory and options for downloading data items individually. Panel 2 shows the plot of monthly precipitation time-series. Panel 3 allows users to visualize time-series of depth to groundwater table at different monitoring wells shown on the map. Panel 4 plots water use time-series estimated from electricity consumption.

Table 4.4 List of variables accessible in Data Portal module

Variable name	Support for download	Support for visualization
Monthly precipitation	Yes	Yes
Annual electricity use	Yes	No
Annual crop area	Yes	Yes
Monthly irrigation water use	Yes	Yes
Monthly groundwater level/depth	Yes	Yes

4.4.3   Architecture of the GTDSS

Figure 4.51 shows the scheme of the Guantao DSS and the data flow between different modules. Specifically, the database holds various types of information ranging from observation data to model parameters used by individual modules. The **Irrigation Calculator**, the **0-D groundwater model (box model)** and the **2-D groundwater**

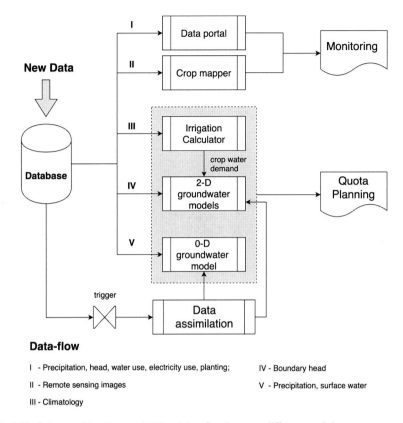

Fig. 4.51 Scheme of the Guantao DSS and data flow between different modules

model are the main modelling components, with which one can run model simulations for given scenarios and assess outcomes (i.e., scenario analysis), or, inversely, specify the desired outcome and explore best strategies to get there (i.e., planning). In addition, the groundwater model parameters and initial conditions of the 2-D groundwater model are updated periodically with the Ensemble Kalman-Filter, which is a modern data assimilation technique to keep a model up to date without manual intervention. The monitoring components consist of the **CropMapper** module for preparing spatial maps of the main irrigated crops in Guantao, and the data portal module for visualizing essential observation data, such as groundwater heads.

References

ADB (2013). Climate Change Adaptation through Groundwater Management of Shanxi Province, People's Republic of China. Asian Development Bank, Loan Project 'Shanxi integrated Agricultural Development Project'. Final Report 0188. PRC ADB Grant Project

Allen RG, Pereira LS, Raes D, Smith M (1998). Crop Evapotranspiration (guidelines for computing crop water requirements). FAO Irrigation and Drainage Paper. http://www.fao.org/3/x0490e/x0490e00.htm

Basak D, Pal S, Patranabis DC (2007) Support vector regression. Neural Inf Process Lett Rev 11:203–224

Chiang WH, Kinzelbach W (2000) 3D-Groundwater modeling with PMWIN—a simulation system for modeling groundwater flow and pollution. Springer, Berlin Heidelberg New York, p 346p

Hadded R, Nouiri I, Alshihabi O, Maßmann J, Huber M, Laghouane A, Yahiaoui H, Tarhouni J (2013) A decision support system to manage the groundwater of the Zeuss Koutine aquifer using the WEAP-MODFLOW framework. Water Resour Manage 27(7):1981–2000. https://doi.org/10.1007/s11269-013-0266-7

Harbaugh AW, Banta ER, Hill MC, McDonald MG (2000) MODFLOW-2000, the U. S. Geological Survey modular groundwater model—user guide to modularization concepts and the groundwater flow process, Open-File Rep. 00-92, U.S. Geol. Survey, Reston, Va

Hendricks Franssen HJ, Kinzelbach W (2008) Real-time groundwater flow modeling with the Ensemble Kalman Filter: Joint estimation of states and parameters and the filter inbreeding problem. Water Resour Res 44:W09408. https://doi.org/10.1029/2007WR006505

Hewitt RJ, Macleod CJA (2017) What do users really need? Participatory development of decision support tools for environmental management based on outcomes. Environments 4(4):88. https://doi.org/10.3390/environments4040088

Kumar S, Godrej AN, Grizzard TJ (2015) A web-based environmental decision support system for legacy models. J Hydroinf 17(6):874–890. https://doi.org/10.2166/hydro.2018.088

Li Y (2020). Towards fast prototyping of cloud-based environmental decision support systems for environmental scientists using R Shiny and Docker. Environ Model Softw 132:0–9https://doi.org/10.1016/j.envsoft.2020.104797

Li N, Kinzelbach W, Li HT, Li WP, Chen F (2019) Decomposition technique for contributions to groundwater heads from inside and outside of an arbitrary boundary: application to Guantao County, North China Plain. Hydrol Earth Syst Sci 23:2823–2840. https://doi.org/10.5194/hess-23-2823-2019

Li N, Kinzelbach W, Li HT, Li WP, Chen F (2021) Improving parameter and state estimation of a hydrological model with the ensemble square root filter. Adv Water Resour 147:103813.https://doi.org/10.1016/j.advwatres.2020.103813

Li Y, Kinzelbach W, Wang H, Ragettli S, Lei ZX, Li WP, Pan H (2021) Management of groundwater overpumping in arid regions: Lessons from Luotuocheng irrigation district, China. (Submitted for publication)

Liu J, Dietz T, Carpenter SR, Alberti M, Folke C, Moran E, Pell AN, Deadman P, Kratz T, Lubchenco J, Ostrom E, Ouyang Z, Provencher W, Redman CL, Schneider SH, Taylor WW (2007) Complexity of coupled human and natural systems. Science 317(5844):1513–1516. https://doi.org/10.1126/science.1144004

Liu Z (2016) Analysis on the reform of water resources management in Minqin county. Gansu Sci Technol 32:10–13 (in Chinese)

Loucks DP (1995) Developing and implementing decision support systems: a critique and a challenge. J Am Water Resour Assoc 31(4):571–582. https://doi.org/10.1111/j.1752-1688.1995.tb03384.x

Matthies M, Giupponi C, Ostendorf B (2005) Environmental decision support systems: current issues, methods and tools. Environ Model Softw 22(2):123–127. https://doi.org/10.1016/j.env soft.2005.09.005

McDonald S, Mohammed IN, Bolten JD, Pulla S, Meechaiya C, Markert A, Nelson EJ, Srinivasan R, Lakshmi V (2019). Web-based decision support system tools: the Soil and Water Assessment Tool Online visualization and analyses (SWATOnline) and NASA earth observation data downloading and reformatting tool (NASAaccess). Environ Model Softw 120:104499. https://doi.org/10.1016/j.envsoft.2019.104499

Milly PCD, Betancourt J, Falkenmark M, Hirsch RM, Kundzewicz ZW, Lettenmaier DP, Stouffer RJ (2008) Stationarity is dead: whither water management? Science 319(5863):573–574. https://doi.org/10.1126/science.1151915

Mir SA, Quadri SMK (2009) Decision support systems: concepts, progress and issues a review. In: Lichtfouse E (ed) Climate change, intercropping, pest control and beneficial microorganisms. Springer, Netherlands, Dordrecht, pp 373–399

Morton MSS (1971) Management decision systems: computer-based support for decision making. Graduate School of Business Administration, Harvard University, Division of Research

Pachauri RK, Allen MR, Barros VR Broome J, Cramer W, Christ R, Church JA, Clarke L, Dahe Q, Dasgupta P, et al. (2014). Climate Change 2014: Synthesis Report. Contribution of Working Groups I, II and III to the Fifth Assessment Report of the Intergovernmental Panel on Climate Change

Quilty J, Adamowski J, Boucher MA (2019) A stochastic data-driven ensemble forecasting framework for water resources: a case study using ensemble members derived from a database of deterministic wavelet-based models. Water Resour Res 55(1):175–202. https://doi.org/10.1029/2018WR023205

Ragettli S, Herberz T, Siegfried T (2018) An unsupervised classification algorithm for multi-temporal irrigated area mapping in Central Asia. Remote Sens 10(11):1823. https://doi.org/10.3390/rs10111823

Sanz D, Calera A, Castano S, Gomez-Alday JJ (2016) Knowledge, participation and transparency in groundwater management. Water Policy 18:111–125. https://doi.org/10.2166/wp.2015.024

Shao H, Yang W, Lindsay J, Liu Y, Yu Z, Oginskyy A (2017) An open source gis-based decision support system for watershed evaluation of best management practices. J Am Water Resour Assoc 53(3):521–531. https://doi.org/10.1111/1752-1688.12521

Swain NR, Latu K, Christensen SD, Jones NL. Nelson EJ, Ame, DP, Williams GP (2015) A review of open source software solutions for developing water resources web applications. Environ Model Softw 67:108–117. https://doi.org/10.1016/j.envsoft.2015.01.014

Walker WE, Lempert RJ, Kwakkel JH (2013) Deep uncertainty. In: Gass SI, Fu MC (eds) Encyclopedia of operations research and management science. Springer, US, pp 395–402

Wang XW, Shao JL, Van Steenbergen F, Zhang QL (2017) Implementing the prepaid smart meter system for irrigated groundwater production in northern china: status and problems. Water 9(6):379. https://doi.org/10.3390/w9060379

Wang L, Kinzelbach W, Yao HX, Steiner J, Wang H (2020) How to meter agricultural pumping at numerous small-scale wells? An indirect monitoring method using electric energy as proxy. Water 12(9):2477. https://doi.org/10.3390/w12092477

Whateley S, Walker JD, Brown C (2015) A web-based screening model for climate risk to water supply systems in the north-eastern United States. Environ Model Softw 73:64–75. https://doi.org/10.1016/j.envsoft.2015.08.001

Yang GY, Gu JF (2008) Handan water resources assessment. Xueyuan Press, Handan (In Chinese)

Yang D, Chen J, Zhou Y, Chen X, Chen X, Cao X (2017) Mapping plastic greenhouse with medium spatial resolution satellite data: Development of a new spectral index. J Photogramm. Remote Sens 128:47–60. https://doi.org/10.1016/j.isprsjprs.2017.03.002

Zoha A, Gluhak A, Imran MA, Rajasegarar S (2012) Non-intrusive load monitoring approaches for disaggregated energy sensing: a survey. Sensors 12:16838–16866. https://doi.org/10.3390/s121216838

Zulkafli Z, Perez K, Vitolo C, Buytaert W, Karpouzoglou T, Dewulf A, De Bièvre B, Clark J, Hannah DM, Shaheed S (2017) User-driven design of decision support systems for polycentric environmental resources management. Environ Model Softw 88:58–73. https://doi.org/10.1016/j.envsoft.2016.10.012

Chapter 5
Way Forward

The combination of fallowing and substituting groundwater by surface water was effective in reducing aquifer depletion in Guantao. The average annual depletion rate after 2014 was about half the value of the pre-project period 2000–2013 and basically limited to the deep aquifer. The goal of closing all deep aquifer wells has only been reached partially, their use being necessary in locations where the shallow aquifer is too saline. Ending unsustainable groundwater use in Guantao and the NCP is feasible with presently available options. It involves a set of measures, large and small, which are all needed. Uncertainty in the future role of groundwater resources is introduced by climate change and socio-economic development. The transition from small family farms to large industrial farms will increase the overall efficiency of farming including water use efficiency. Population decrease and diet change both affect food demand and thus future agricultural structure and irrigation demand. New cultivars may increase yield without increasing water needs. While the current over-pumping control heavily depends on surface water imports from the South, climate change induces uncertainty by both reducing water availability in the South and increasing water deficits of double cropping in the NCP.

5.1 Recommendations for Guantao

The situation of groundwater over-pumping in Guantao has improved since 2014. The groundwater gap of about 37 Mio. m^3/year before 2014 has been reduced to almost one half during the project. The shallow aquifer is basically in equilibrium between recharge and discharge. The remaining task is stopping the decline of piezometric levels in the deep aquifer, which implies the abandoning of all deep wells. The corresponding reduction of pumping from the deep aquifer of approximately 20 Mio. m^3/year can be implemented either by decreasing demand or increasing supply. Further fallowing of winter wheat, making better use of present surface water

W. Kinzelbach et al., *Groundwater Overexploitation in the North China Plain: A path to Sustainability*, Springer Water, https://doi.org/10.1007/978-981-16-5843-3_5

imports, adding more reliable surface water imports at time of use, and saving water through agricultural techniques certainly have the necessary potential (see analysis in Box 5.1).

In its work in Guantao, the World Bank's GEF project (Foster and Garduño 2004) stressed that only reduction of consumptive water use (i.e. transpiration and evaporation) is real water saving. This remains of course true, but looking at saving of deep aquifer water, the statement has to be modified. As there is no irrigation back flow or any other seepage back to the deep aquifer, any reduction of pumping is 100% saving of deep aquifer water, irrespective of whether it turns into evapotranspiration, seepage to the shallow aquifer, or drainage outflow.

Main aquifer rehabilitation measures, their remaining potentials in Guantao and unit costs

Reduction of groundwater abstraction can be achieved by supply side measures, adding new water resources, and by demand side measures, reducing abstractions by water saving. Costs of both can be expressed by CNY per cubic meter supplied or saved. Demand side measures (darker shading in the table) are subsidy driven. Supply side measures consist of water imports from reservoirs, the Yellow River or - via the SNWT - from the Yangtze River Basin.

Measure	Remaining Potential in Guantao (Mio. m^3/yr)	Cost/Subsidy (CNY/m^3)
Traditional water saving* (small fields and low pressured pipe irrigation)	0.4	1.6
Highly efficient water saving* (drip irrigation in green houses)	0.7	2.5
Highly efficient water saving* (sprinklers)	0.8	2.5
Mulching*	0.2	1.0
Conservation tillage* (maize)	< 1	<0.5
Conservation tillage* (wheat)	0.6	1.3
Rain fed cropping**	0.9	4.5
Water saving cultivars***	4	1.9
Fallowing of winter wheat**	10	3.0
Yellow River and reservoir water (No increase, improvement of efficiency in use of existing imports only)	< 15	0.4
SNWT Central Route (Replace remaining deep aquifer pumping for drinking water supply)	4	2.5
SNWT Eastern Route	(Extension to Hebei under construction)	< 1

*   These measures contribute to water saving by reducing unproductive evaporation. If the irrigation water is taken from the deep aquifer, all reduction in pumping contributes to real water saving.

** These measures save groundwater by not pumping at all.

*** These measures save groundwater by increasing WUE of crops.

Main sources for data: a) Work experience summary of groundwater overexploitation control in Hebei Province. Jianghe Conservancy and Hydropower Consulting Center (2018). b) Self-assessment report on groundwater overexploitation control in Handan prefecture. Handan Municipal People's Government (2019-2020). c) The third-party evaluation report on the pilot project of comprehensive groundwater overexploitation control in Hebei Province (MWR/IWHR 2014). (All in Chinese).

Box 5.1: Aquifer rehabilitation measures: remaining potential in Guantao and specific costs. The box shows two prominent items, improved used of water imports from the south and increased fallowing of winter wheat, which taken together are sufficient to close a gap of 20 Mio. m³/year. They are followed in size by water saving through new cultivars (4 Mio. m³/year) and numerous "small" measures of water conservation, also adding up to a potential of about 4 Mio. m³/year. Deep aquifer pumping by households and industry should be completely substituted by additional imports of 4 Mio. m³/year of SNWT water. All measures together (30 Mio. m³/year) should be sufficient to fill the deep aquifer gap in the coming years provided the continued commitment of the administration.

The success of the last 7 years is mainly due to the fallowing measures and the import of surface water. It does not mean that no more management of the shallow aquifer is necessary. On the contrary, only from this situation as a starting point the storage of the shallow aquifer can be managed actively, hopefully increased, as a storage device to overcome future drought years. The system in place is also important to react to more systematic changes in the future, be it a change in crop water demand due to rising temperatures or changes in recharge due to changes in rainfall and rainfall patterns. The decision support system can keep track and help in designing the right answers in an adaptive management approach charted out by the red line concept.

The metering of water use and the collection of fees either for amounts overstepping quota or for every cubic meter of water pumped certainly serves the purpose of ensuring general discipline in water use. It is however only justified if the fees lead to a reduction of pumping. This can happen either directly or indirectly by using the collected fees to subsidize water saving measures.

Metering by electricity consumption is practicable and makes economic sense. It should be developed further with the electricity company using their daily available "big data". It would be ideal if the water fee could be collected automatically together with the electricity fee saving the effort for a second fee collection system.

Much has been achieved in the last years and a combination of crop system change and South-North Water Transfer (SNWT) clearly has sufficient potential to stop aquifer depletion, not only in Guantao but in the whole of the NCP. But stopping the depletion is not enough. Even if a cone of depression is stabilized by surface water imports and a hydraulic quasi-steady state can be reached, such a closed system will salinize with time. Only if drainage outflows to the sea via canals, streams, rivers and the aquifer are restored one can truly speak of a sustainable solution. In future, the focus should turn to groundwater quality. The salinity of pumped water should be monitored at least once per year. The increased agricultural pumping in the deep aquifer of the last few years is probably already related to an increase in salinity in the shallow aquifer's cones of depression. While the end of over-pumping is within reach, the salinity problem will not go away and requires further careful management.

5.2   Agriculture and Future Groundwater Management in the NCP

Groundwater depletion in the NCP was a consequence of the governments' will to battle famine in the early 1980s. Hebei Province played an important role in feeding the country since then. It increased its yield by 2–3 times over the past 3 decades. The intensification of food production coincided with a time of decreasing rainfall, which magnified the consequences regarding groundwater levels.

The centralized governmental structure in China usually allows the Chinese leaders to mobilize a vast amount of resources to pursue a goal or solve any problem in a determined manner once an issue is confirmed as extremely important. If a problem is perceived as in conflict with other objectives of interest, any decision is essentially a trade-off. This is the case for groundwater management in the NCP because mitigating groundwater over-pumping may hurt national food security and farmers' income, which are both considered at least as important as environmental protection, if not more important. In 2021, the agricultural department again received the task of increasing crop production (http://www.xinhuanet.com/2021-01/03/c_1 126940702.htm).

Population will peak between 2025 and 2030 at about 1.5 billion (Chen et al. 2020), who want to have an ever-richer diet with more meat, vegetables, and fruit. Climate change will decrease the yield of some of today's cultivars due to higher temperature and increase the water demand of crops due to higher ET (see Sect. 5.5). Water scarcity is exacerbated by deteriorating soil and water quality. Climate change may also put a question mark to the water transfers from the South, which must maintain a reserve for its own growth. So, will the solutions envisaged today still hold in 2050 or does it take more? To approach this question, a look at some trends of today is helpful.

5.2.1   Transition from Smallholder Agriculture to Larger Farms

The first trend is the transition from the present smallholder agriculture to modernized larger-scale farming in the NCP. The smallholder family farms with a cropping area of a third of a hectare, are economically unattractive and have no future. The annual income from a winter wheat and a summer maize crop on that area is less than 3000 CNY per person , which compares unfavorably with the average annual income in China of about 30'000 CNY per person. Young people leave the countryside and look for jobs in the cities. Their support of their parents is usually more significant than the parents' own income from farming. Most people working in the fields are above 45 and the question must be asked, who will produce China's food in 2050 (see Box 5.2). The situation is expected to change dramatically in the next decade(s) due to the fast change in rural population structure. With it, agricultural production activities and organization forms will change substantially too, from smallholder farming to larger farms through land aggregation.

Take Switzerland for example: For an economically viable farm of 4 persons in Switzerland a farm area of about 100 ha (300 times the area of a Chinese family farm or almost the size of the whole cropland of a village in the NCP) is necessary. It is typically managed by the family plus one additional farm hand, which is feasible due to a high degree of mechanization, applying modern technology both for remotely controllable hardware, including machinery and irrigation equipment, and for information technology harnessing information on plant health, weather and markets for example.

Chinese farming is developing in the same direction. Farm sizes in the NCP are already increasing. Big entrepreneurs lease land from individual farmers, sometimes of whole villages, to practice a larger scale and more efficient style of farming. This opens up new opportunities also for increased efficiency in water use. However, it is unclear how grain production can eventually become profitable for big farms relying on land leased from small farm households. With the current leasing cost (800–1000 CNY/mu/year), planting grain crops (wheat and maize) is not sufficiently profitable to attract farming enterprises to take over. Agricultural land ownership and the right to trade ownership in a more permanent way will become more and more a critical issue in this transition process.

Using engineering and agronomic approaches, larger farm units can fully exploit the potential for improving water use efficiency beyond today's level. While highly efficient center-pivot sprinklers are neither applicable nor affordable for today's small family farms, they can unfold their full water saving potential on farms of at least 100 ha in size. A large farm can also afford to fallow a certain percentage of its area in order to let the soil rest or to plant green manure without having to be pushed by government. Today's pioneers of big farming go into niche products such as plant seeds, health foods, medicinal plants, or farm tourism to optimize income. At the next stage, grain production should be considered for scale-up to a more efficient and profitable operation.

Greenhouses are the most profitable production units, even at the scale of a family farm. When equipped with soil moisture sensors and automatic drip- or micro-sprinkling equipment they can save significant amounts of water compared to the traditional green houses. At the same time, they can save labor, which meanwhile has become expensive in China. In the US, vertical farms are quickly expanding. They produce vegetables in hydroponic cultures close to the consumer. Their water use is minimal as most of the transpired water can be recycled. And they do not need any agrochemicals except nutrients. Economic viability relies on the consumers' demand for healthy food. China's urban elite is also conscious of food quality, which shows in collective ordering or "crowd farming" of farm products from trusted producers in the periphery.

Bigger farms allow a more scientific approach to crop production, often termed "precision agriculture", which benefits the quality of the produce as well as the environment. Drones already today apply crop protection products. In future they will be able to deposit those products on plants in need only. Similar technology can improve fertilizer application to conserve nutrients and reach more uniform production over the area.

The smart water meters, which have been successfully applied in China's West will also be sustainable in the NCP if pumping in big farms is concentrated on fewer wells. Management of water quota by a county's water resources administration will become feasible when it needs to address only a small number of large farms instead of thousands of small farms. Already during the project, the biggest farm of Guantao enthusiastically volunteered to cooperate with our metering experiment. They want to save water—and thus energy—as they are much more conscious of cost-efficiency than the smallholder farmers. Smart devices such as smart water meters with data transmission to a central server and eventually a feedback from that server to the device, mark the beginning of the Internet of Things (IoT) in agriculture.

Information technology and the full use of available information have to play a bigger role in farming. Alibaba launched the "ET Agricultural Brain". This digital tool enables farmers to raise their crop yields and income by leveraging big data. The app allows farmers to digitally record information about their yields in order to optimize their entire production cycle, raising efficiency and capacity. It was preceded by similar efforts in North America where for example Climate Corporation analyses 50 terabytes of weather data every day and sends the information to users via its Software-as-a-Service platform Climate.com. The corporation's goal is to leverage data to help farmers deal with the increasingly volatile weather caused by climate change.

Agricultural research in China is quite advanced, but it is slow in trickling down into practice. A recent example shows how it can be made available to farmers faster: Through a national campaign, about 20.9 million farmers adopted the recommendations of a study, which increased productivity and reduced environmental impacts. As a result of the intervention, farmers were together US$12.2 billion better off through savings in fertilizer and increase in yield (Cui et al. 2018). The new type of agriculture requires human resources with the relevant education. To feed China's population in 2050, a scientifically educated workforce with an agricultural background will be vital (Tso 2004).

Increase of production per ha does not automatically mean increase of production per cubic meter of water, i.e. more "crop per drop". Therefore, optimization of production must go in parallel with other options of reducing water use. Agrotechnology has still some other methods in store, which under the name of "new green revolution" will help to optimize the combined efficiency of water and nitrogen use (see Sect. 5.3).

5.2.2   Consumer Behavior and Food Demand

Water saving must include the **reduction of food waste** post-harvest, in retail and on the level of the consumer. In recent years Chinese propaganda campaigns focused on the reduction of food waste in restaurants. In Europe initiatives push smart logistics for internet-based distribution of food close to expiration date. Food saved is water saved.

Future water demand is closely related to the diet people desire (McLaughlin and Kinzelbach 2015). Even a stagnating population will further increase water demand if the present trend towards increasing consumption of meat and milk products continues. In the past 30 years China's meat consumption has grown 2.5 fold with a population increase of about 15%, and the trend towards more meat continues, with values admittedly still low in comparison to many western countries (see Box 5.3)

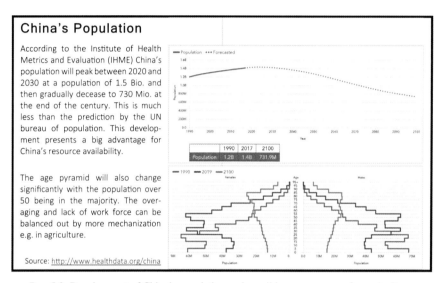

China's Population

According to the Institute of Health Metrics and Evaluation (IHME) China's population will peak between 2020 and 2030 at a population of 1.5 Bio. and then gradually decease to 730 Mio. at the end of the century. This is much less than the prediction by the UN bureau of population. This development presents a big advantage for China's resource availability.

	1990	2017	2100
Population	1.2B	1.4B	731.9M

The age pyramid will also change significantly with the population over 50 being in the majority. The over-aging and lack of work force can be balanced out by more mechanization e.g. in agriculture.

Source: http://www.healthdata.org/china

Box 5.2: Development of China's population and possible consequences for agriculture.

.

However, the trend in meat consumption could change over health concerns. Young people in the west turn to vegetarianism. While the number of vegetarians in Germany was estimated to be 0.6% in 1983, it increased to about 10% of the population in 2016, with typical vegetarians being between 19 and 29 years old (Mensink et al. 2016). The market adapted to this dietary shift and the associated new demand. Sales of meat substitutes are increasing rapidly, creating a new billion-Dollar market.

The **diet** plays an important part in determining the water demand in agriculture. Of course, its meat component impacts in the first place the production of feed grain, which grows in summer and presents a smaller potential for water saving than wheat. But **substitutes** for wheat could also be considered. Potatoes yield the largest number of calories with the least input of water per unit area. Of course, dietary changes need time for people to get accustomed to. But did China not get accustomed to wheat after centuries of millet? Potatoes, which were unknown before Columbus, are the 4th staple food crop (after rice, wheat and maize) in China since 2015 (Xinhua 2015).

Possibly China may also get used to Quinoa, which is advocated by the UN in its efforts to enhance the globe's food security while saving water (FAO 2011).

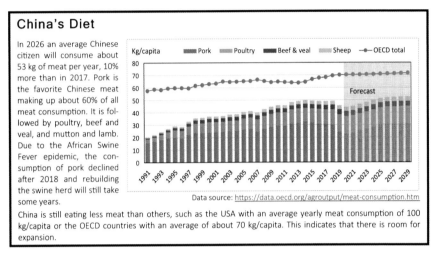

China's Diet

In 2026 an average Chinese citizen will consume about 53 kg of meat per year, 10% more than in 2017. Pork is the favorite Chinese meat making up about 60% of all meat consumption. It is followed by poultry, beef and veal, and mutton and lamb. Due to the African Swine Fever epidemic, the consumption of pork declined after 2018 and rebuilding the swine herd will still take some years.

Data source: https://data.oecd.org/agroutput/meat-consumption.htm

China is still eating less meat than others, such as the USA with an average yearly meat consumption of 100 kg/capita or the OECD countries with an average of about 70 kg/capita. This indicates that there is room for expansion.

Box 5.3: Development of China's meat consumption

5.3   Drought Resistant and High Yield Wheat Varieties

Managing groundwater sustainably depends to a high degree on agricultural strategies adopted in the NCP. Pursuing high yield to feed people was the major agricultural target pursued during the past 4 to 5 decades. The NCP played an important role in improving grain production during this period, when numerous high yield varieties were bred and widely cultivated. The yields of wheat and maize increased considerably from 1980 to 2010 (Fig. 5.1). They can reach 9 t/ha for wheat and 10.5 t/ha for maize today. This is much higher than the respective national average yields of 6.2 t/ha and 5.4 t/ha (2019 data) and opens up a yield gap. While the development of crop varieties has helped to achieve the goal of self-sufficiency in agricultural goods, the yield gap still presents a reserve to be activated.

The achievements in grain production were accompanied by a tremendous increase in fertilizer and irrigation water inputs, with the well-known consequences for groundwater depletion. While over the last 40 years yield increase was the primary objective, now the reduction of agricultural intensity has become an essential option to mitigate groundwater level decline.

Increasing attention has been given to the applications of biologic approaches, such as breeding of drought resistant varieties, or genetic modification and gene editing to largely improve photosynthetic efficiency. According to 30-year field experiments at the Center for Agricultural Resources Research, Institute of Genetics

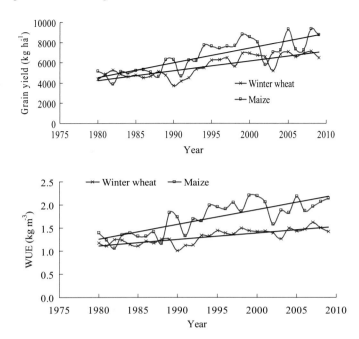

Fig. 5.1 Changes of grain yield and WUE for wheat and maize in NCP from 1980 to 2010 (Zhang et al. 2011)

and Developmental Biology, CAS, a wheat variety with high water use efficiency could save 30–50 mm of irrigation water compared to the normal varieties planted in NCP (Fang et al. 2017) with almost the same level of yield. The WUE could be improved by 21.6% maximally by adoption of highly water productive varieties for wheat. On average, adopting high WUE varieties could achieve an increase of grain yield by about 1% and an increase in WUE by 0.5% annually (Zhang et al. 2010).

A recent study revealed that nitrogen status affects rice chromatin function through the modification of histones, and some specific genes regulating tillering. The transcription factor NGR5 mediates hormone signaling and increased NGR5 levels can drive increases in rice tillering and yield without increasing nitrogen fertilization requirements (Wu et al. 2020). This kind of breakthrough in genetic or biological sciences can be expected to greatly improve water and nitrogen use efficiencies of grain crops. Therefore, another "green revolution" through genetic editing technologies may happen in the near future. It should result in new drought resistant crop varieties and varieties with higher photosynthetic ability, using less water and nitrogen fertilizer.

5.4   Importing Surface Water: Benefit and Risk

For regions lacking large rivers while suffering from severe water shortage, groundwater overdraft seems to be inevitable and importing surface water from outside of the region is the most direct and effective way to alleviate the decline of groundwater levels. Based on the experience from groundwater overdraft control in the NCP so far, importing surface water should not only increase the availability of water but also the efficiency of its use. Present use of surface water imports is extremely inefficient as the timing of water transfers depends on flows in the upstream, which are not synchronous with the irrigation calendar. Storage capacity available in the plain area is by far not sufficient to allow more efficient use of the imports. Improvement requires further updating of water infrastructure including the water system's connectivity and maintenance, e.g. by dredging river channels, renovating pools and ponds, as well as using imported water for MAR.

Import of surface water has been planned, designed, and implemented rather well compared to other methods of groundwater over-pumping control so far. Although it seems irreplaceable in the struggle to end groundwater depletion in the NCP, it is still facing considerable risks and uncertainties.

The surface water quota from the Yellow River will still be strictly limited in the near future. Due to constraints on water quota, after three years of groundwater over-pumping governance, almost all cities and counties have used up their quota of Yellow River water. However, some counties' task of cutting down on groundwater extraction has not been achieved. In Hebei Province Yellow River water has mainly been used for agricultural irrigation. If no further imports are possible, groundwater over-pumping governance can only rely on optimization of the cropping structure, which depends heavily on government subsidies. With the decline of these, the further promotion of over-pumping control action in these mainly rural areas remains challenging.

The high cost of water supply from the SNWT project brings risk and uncertainty to the groundwater over-pumping control system. One important reason is that Hebei provincial government can only spend a small portion of its whole budget for projects supporting the SNWT. 78% of the investment relies on bank loans. The construction of water treatment plants and connecting water supply pipelines is mainly supported by financing through non-governmental investors. These investors expect high returns on their investments. The above two factors result in high water supply and transfer costs for the SNWT. The water price at the entrance of the main canal of the SNWT Project into Hebei Province is 0.97 CNY/m^3, the price of water entering the surface water treatment plant is 2.76 CNY/m^3, and the water price for the consumers is 7.43 CNY/m^3, which is 24% higher than in Beijing, and 34% higher than in Tianjin. It amounts to 155% of the current average water price in urban areas, and to 210% of the current county average water price of Hebei Province.

Due to its high price, water of the SNWT Project is not affordable for agricultural and ecological purposes. Even redundant SNWT water is too expensive to replace groundwater in agricultural irrigation and preserve the groundwater system.

If the present water price of the SNWT's Central Route of 2.5 CNY/m^3 is assumed, the water needed for irrigation of winter wheat would cost about 6′000 CNY/ha. This amount is close to the present input cost for planting one ha of winter wheat and would make wheat planting with SNWT water economically infeasible. Today, the irrigation cost of wheat with groundwater (including maintenance and labor) is about 1000 CNY/ha or 1/6th of the total input cost. The cost of water of the Eastern Route will be less than that of the Central Route, but it cannot be cheaper than groundwater as it has to be pumped up by about 100 m from the Yangtze River, about twice the lift presently necessary for deep aquifer groundwater in the NCP. If the management of all irrigation water would be put into the responsibility of one big company, they could make an average price, which would dilute the high marginal cost of the transfer water to an acceptable prize for food production.

5.5   Climate Change in the NCP

Climate change influences the fate of groundwater resources in the NCP in many ways. Precipitation determines not only groundwater recharge. It also determines the need of supplementary irrigation and therefore groundwater abstraction. Its distribution in time, extreme rain events or drought events can be as important for agricultural production as the annual total precipitation. Temperature, wind speed and radiation influence the potential evaporation, which finally determines the evapotranspiration potential of the crops and thus the water requirements. Finally, climate change may affect the flow of rivers, which are now the source of water transfers to the NCP.

The NCP has seen climate change in the past. Observations since the mid-twentieth century show that it has become warmer and drier. Most observed downward trends in annual precipitation amount were statistically significant (Chen et al. 2010; Fu et al. 2009; Liu et al. 2005; Sun et al. 2017; Wang et al. 2012). Summer precipitation, which accounts for 50–75% of the total annual rainfall in NCP, also has decreased significantly (Fan et al. 2012; Sun et al. 2020; Wang et al. 2012; Ye 2014), while precipitation in other seasons has shown contrasting trends (Fan et al. 2012; Sun et al. 2020). The decrease in rainfall led to a decrease in groundwater recharge. Significant increase was not only observed in the mean annual temperature but also in temperature during both wheat and maize seasons (Chen et al. 2010; Liu et al. 2014; Zhang et al. 2015). Surprisingly, a long-term decreasing trend has been observed in actual evapotranspiration, which can be explained by decreasing trends in precipitation, sunshine duration and wind speed which overcompensated the increase in ET due to temperature rise alone (Cao et al. 2014; Chen et al. 2010; Fan et al. 2012; Liu et al. 2014; Song et al. 2010). A recent study based on remote sensing and physical modeling found that actual evapotranspiration increased slightly since 2000, but concluded that the contribution of climate change was less than that of human activities (Chen et al. 2017). The result agrees with the measurement data of Luancheng Agro-Eco-Experimental Station of the Chinese Academy of Sciences showing that the actual seasonal ET of winter wheat and summer maize under well-watered conditions gradually increased from the 1980s to the 2000s possibly caused by the increase in leaf stomatal conductance associated with the introduction of new cultivars (Zhang et al. 2011). With global warming, more crops can be grown in Northeast China, which has better rainfall, and the recent decades already have shown some shift in production from the NCP towards the Northeast, reducing the pressure on the NCP.

The Fifth Assessment Report by the Intergovernmental Panel on Climate Change (Pachauri et al. 2014) provides the analysis of model outputs from the fifth phase of the Coupled Model Intercomparison Project (CMIP5). CMIP5 produced a state-of-the-art multi-model dataset designed to advance our knowledge of climate variability and climate change including "long term" simulations of twentieth-century climate and projections for the twenty-first century and beyond for a number of Representative Concentration Pathways (RCPs) for atmospheric greenhouse gas (GHG) concentrations (Van Vuuren et al. 2011). Annual averages of temperature and precipitation for two RCPS and a time interval at the end of the century are shown in Fig. 5.2. Looking at Eastern China for the more pessimistic RCP8.5, one can find a strong increase in yearly average temperature (5–7 °C) and an increase in annual rainfall (10–20%) compared to the baseline of 1986–2005. For a lower emission scenario, annual rainfall hardly changes, while average temperature increase is about 1 °C.

To adapt these global projections to regions, a process called downscaling is required to obtain a higher spatial resolution suited for the scale of the region. The problem with the projections is their uncertainty. While different models usually agree well as far as the projected temperature rise is concerned, predictions of rainfall are extremely uncertain reaching from increase to decrease of rainfall. This is also true for the NCP, where the ensemble of climate models shows discrepancies among GCMs and downscaling methods, concerning the projected mean and extremes of precipitation. Some studies show a likely increase in annual precipitation but at varying rates (Fu et al. 2009; Tao and Zhang 2013), while others project a decline in annual precipitation (Liu et al. 2013). Being influenced by several climate variables, evapotranspiration estimates turn out to be very uncertain, making it difficult to draw conclusions about future trends. The projected changes of evapotranspiration during the wheat or wheat–maize growth period in the NCP could either increase or decrease within a range of around ± 10% depending on the GCM and greenhouse gas emission scenario employed (Guo et al., 2010; Lv et al. 2013; Mo et al. 2013; Tao and Zhang 2013).

We can distinguish direct and indirect impacts of climate change on groundwater resources. Changes in the amount, intensity and frequency of precipitation directly determine groundwater recharge and indirectly influence the demand of groundwater pumping for irrigation. The direct impacts of changes in precipitation on groundwater recharge have been found to contribute less to groundwater resources than the changes in pumping, not only in the NCP but also in other regions (Larocque et al. 2019; Li et al. 2014). However, it should be noted that studies on climate change assessments usually ignore the indirect impacts of precipitation on groundwater pumping, and

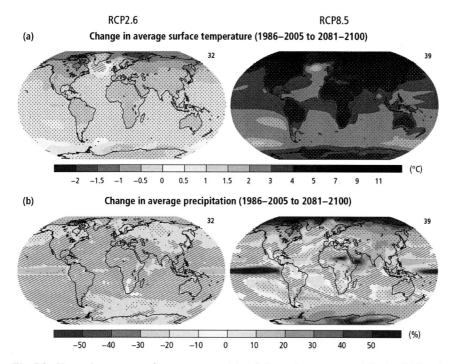

Fig. 5.2 Change in average surface temperature (a) and change in average precipitation (b) based on multi-model mean projections for 2081–2100 relative to 1986–2005 under the RCP2.6 (left) and RCP8.5 (right) scenarios (Pachauri et al. 2014). The number of models used to calculate the multi-model mean is indicated in the upper right corner of each panel. Stippling (i.e., dots) shows regions where the projected change is large compared to natural internal variability and where at least 90% of models agree on the sign of change. Hatching (i.e., diagonal lines) shows regions where the projected change is less than one standard deviation of the natural internal variability

these indirect impacts are usually connected to human activities. Similarly, evapotranspiration also directly influences groundwater recharge and indirectly determines groundwater pumping by crop water demand. Previous studies in the NCP focused on the influence of projected evapotranspiration on future crop water demand and crop yield (Tao and Zhang 2013; Xiao et al. 2020; Xiao and Tao 2016). Hardly any model takes into account the decrease of plant transpiration due to the impact of CO_2 on stomatal conduction, which also modifies estimates of irrigation water demand (Guo et al. 2010).

While an increase in transpiration will decrease recharge and instigate more pumping due to higher plant water demand, there are three mechanisms, which in other regions of the world have shown to increase groundwater recharge: The first is the increase in extremes such as strong rainfall events (Fischer and Knutti 2019). Recharge is not a linear function of rainfall. While small rain events may give zero recharge, large events lead to over-proportional recharge. The second is the decrease of average wind with global warming. As reported above, this trend has been observed in China over the last decades. If it continues, it will reduce the expected increase in evapotranspiration due to temperature increase. The third mechanism is due to the increase of winter precipitation also quoted from literature above. At warmer winter temperatures, the soil freezing depth will be reduced and more recharge will occur in the cold season when evaporation is low, especially when there is thawing snow.

To get a coherent picture for the trends in evapotranspiration and the resulting crop water demand, we did our own assessment. The results presented in Box 5.4 show that overall, the crop water deficit will increase by mid-century and increase even further by late century. The compensation by decreasing wind speed as in the past decades is not seen anymore. That means the second mechanism mentioned above will most probably not be available.

Given all the interactions and assuming that total annual precipitation will not decrease, recharge will most probably not change much compared to today. But, due to more extremes both wet and dry, the use of the aquifer as a buffering device is of growing importance in the future.

The water resources availability in the NCP is not only determined by the development in the NCP itself. As the North in recent years relies increasingly on water imports from the South, the influence of climate change on the water availability in the catchment of the Han River and the Yangtze River in general is also a relevant influence factor.

Assessment of climate change impact on crop water deficit

The assessment compares the crop water deficit under present conditions (computed from ERA5-reanalysis data) with projected conditions at mid and late century obtained from the CORDEX project. Only the two CORDEX experiments were chosen which perform well in a comparison for 2000-2013 (see table).

Experiment ID	GCM	RCM	Period	Scenario experiment
2	ICHEC-EC-EARTH	CLMcom-CCLM5-0-2	2035-2064	RCP4.5
4	MOHC-HadGEM2-ES	CLMcom-CCLM5-0-2	2070-2099	RCP4.5

Note: GCM-global climate model; RCM-regional climate model; RCP-representative concentration pathway.

Workflow:

1. Retrieval of projection data from CORDEX project for East Asia region and extraction of gridded data points covering Guantao region only.
2. Quality control by comparing raw projection data with ERA5 reanalysis data as a proxy of gage measurements over the control period (2000-2013).
3. Computation of monthly Change Factors (CFs) for main climate variables for mid-century (2035-2064) and late century (2070-2099), respectively. Those climate variables include precipitation, min/max air temperature at 2 meter above ground, and cloud cover. The CFs are computed as difference for temperature, and as ratio for precipitation as well as cloud cover between future climate and control period.
4. Simulation of synthetic time series with the weather generator using CFs for Guantao (Peleg et al. 2017).

A water deficit indicator was computed as the difference between ET and precipitation. Presently July is the only month when rainfall is more than sufficient to cover crop evapotranspiration, while all remaining months tend to face water deficit, especially in spring season. However, this surplus is expected to decrease by mid-century, and even more so by late century, due to the increase of crop ET in July. Changes in wind speed and radiation do not compensate the increase of ET due to rising temperatures. Guantao will be facing a more severe water deficit than today.

Estimated changes of water deficit under climate projections.

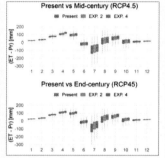

Box 5.4: Influence of climate change on crop water deficit

References

Cao L, Zhang Y, Shi Y (2011) Climate change effect on hydrological processes over the Yangtze River basin. Quat. Int. 244:202–210. https://doi.org/10.1016/j.quaint.2011.01.004

Cao G, Han D, Song X (2014) Evaluating actual evapotranspiration and impacts of groundwater storage change in the North China Plain. Hydrol. Processes 28:1797–1808. https://doi.org/10.1002/hyp.9732

Chen C, Wang E, Yu Q, Zhang Y (2010) Quantifying the effects of climate trends in the past 43 years (1961–2003) on crop growth and water demand in the North China Plain. Clim Change 100:559–578. https://doi.org/10.1007/s10584-009-9690-3

Chen X, Mo X, Hu S, Liu S (2017) Contributions of climate change and human activities to ET and GPP trends over North China Plain from 2000 to 2014. J Geog Sci 27:661–680. https://doi.org/10.1007/s11442-017-1399-z

Chen Y, Guo F, Wang J et al (2020) Provincial and gridded population projection for China under shared socioeconomic pathways from 2010 to 2100. Sci Data 7:83. https://doi.org/10.1038/s41 597-020-0421-y

Cui Z et al (2018) Pursuing sustainable productivity with millions of smallholder farmers. Nature 555:363–366. https://doi.org/10.1038/nature25785

Fan L, Lu C, Yang B, Chen Z (2012) Long-term trends of precipitation in the North China Plain. J Geog Sci 22:989–1001. https://doi.org/10.1007/s11442-012-0978-2

Fang Q, Zhang XY, Chen SY, Shao LW, Sun HY (2017) Selecting traits to increase winter wheat yield under climate change in the North China Plain. Field Crops Research 207:30–41. https:// doi.org/10.1016/j.fcr.2017.03.005

FAO (2011). Quinoa: An ancient crop to contribute to world food security. Technical Report of the FAO Regional Office for Latin America and the Caribbean, 55 p. http://www.fao.org/3/aq287e/ aq287e.pdf

Fischer EM, Knutti R (2016) Observed heavy precipitation increase confirms theory and early models. Nature Clim. Change 6:986–991. https://doi.org/10.1038/nclimate3110

Foster S, Garduño H (2004). China: Towards Sustainable Groundwater Resource Use for Irrigated Agriculture on the North China Plain. World Bank. Sustainable Groundwater Management: Lessons from Practice. GW-MATE Case Profile Collection Number 8

Fu G, Charles SP, Yu J, Liu C (2009) Decadal climatic variability, trends, and future scenarios for the North China Plain. J Clim 22:2111–2123. https://doi.org/10.1175/2008JCLI2605.1

Guo R, Lin Z, Mo X, Yang C (2010) Responses of crop yield and water use efficiency to climate change in the North China Plain. Agric Water Manage 97:1185–1194. https://doi.org/10.1016/j. agwat.2009.07.006

Handan Municipal People's Government (2020). Self-assessment report on groundwater overexploitation control in Handan prefecture (in Chinese).

Jianghe Conservancy and Hydropower Consulting Center (2018). Work experience summary of groundwater overexploitation control in Hebei Province (in Chinese)

Larocque M, Levison J, Martin A, Chaumont D (2019) A review of simulated climate change impacts on groundwater resources in Eastern Canada. Canadian Water Resources Journal/revue Canadienne Des Ressources Hydriques 44:22–41. https://doi.org/10.1080/07011784.2018.150 3066

Li X, L, G, Zhang Y, (2014) Identifying major factors affecting groundwater change in the North China Plain with grey relational analysis. Water 6:1581–1600. https://doi.org/10.3390/w6061581

Liu B, Xu M, Henderson M, Qi Y (2005). Observed trends of precipitation amount, frequency, and intensity in China, 1960–2000. Journal of Geophysical Research: Atmospheres 110https://doi. org/10.1029/2004JD004864

Liu W, Fu G, Liu C, Song X, Ouyang R (2013) Projection of future rainfall for the North China Plain using two statistical downscaling models and its hydrological implications. Stochastic Environ. Res. Risk Assess. 27:1783–1797. https://doi.org/10.1007/s00477-013-0714-1

Liu Y, Yang X, Wang E, Xue C (2014) Climate and crop yields impacted by ENSO episodes on the North China Plain: 1956–2006. Reg Environ Change 14:49–59. https://doi.org/10.1007/s10113-013-0455-1

Lv Z, Liu X, Cao W, Zhu Y (2013) Climate change impacts on regional winter wheat production in main wheat production regions of China. Agric for Meteorol 171–172:234–248. https://doi.org/ 10.1016/j.agrformet.2012.12.008

McLaughlin D, Kinzelbach W (2015) Food Security and Sustainable Resource Management. Water Resour Res 51(7):4966–4985. https://doi.org/10.1002/2015WR017053

Mensink GBM, Lage Barbosa C, Brettschneider AK (2016). Verbreitung der vegetarischen Ernährungsweise in Deutschland. Journal of Health Monitoring 1(2). https://doi.org/10.17886/ RKI-GBE-2016-033

Mo X, Guo R, Liu S, Lin Z, Hu S (2013) Impacts of climate change on crop evapotranspiration with ensemble GCM projections in the North China Plain. Clim Change 120:299–312. https:// doi.org/10.1007/s10584-013-0823-3

MWR/GIWP (2019) Comprehensive Control of Groundwater Overdraft in Hebei Province. Report of GIWP (in Chinese)

MWR/IWHR (2014). The third-party evaluation report on the pilot project of comprehensive groundwater overexploitation control in Hebei Province. Report of IWHR (in Chinese)

Pachauri RK, Allen MR, Barros VR, Broome J, Cramer W, Christ R, Church JA, Clarke L, Dahe Q, Dasgupta P, et al. (2014). Climate Change 2014: Synthesis Report. Contribution of Working Groups I, II and III to the Fifth Assessment Report of the Intergovernmental Panel on Climate Change.

Peleg N, Fatichi S, Paschalis A, Molnar P, Burlando P (2017) An advanced stochastic weather generator for simulating 2-D high-resolution climate variables: AWE-GEN-2d. J Adv Model Earth Syst 9:1595–1627. https://doi.org/10.1002/2016MS000854

Song Z, Zhang H, Snyder RL, Anderson F, Chen F (2010) Distribution and trends in reference evapotranspiration in the North China Plain. J Irrig Drain Eng 136:240–247. https://doi.org/10.1061/(ASCE)IR.1943-4774.0000175

Sun J, Lei X, Tian Y, Liao W, Wang Y (2013) Hydrological impacts of climate change in the upper reaches of the Yangtze River Basin. Quat Int 304:62–74. https://doi.org/10.1016/j.quaint.2013.02.038

Sun J, Wang X, Shahid S (2020) Precipitation and runoff variation characteristics in typical regions of North China Plain: a case study of Hengshui City. Theor Appl Climatol 142:971–985. https://doi.org/10.1007/s00704-020-03344-8

Sun Q, Miao C, Duan Q (2017) Changes in the spatial heterogeneity and annual distribution of observed precipitation across China. J Clim 30:9399–9416. https://doi.org/10.1175/JCLI-D-17-0045.1

Tao F, Zhang Z (2013) Climate change, wheat productivity and water use in the North China Plain: a new super-ensemble-based probabilistic projection. Agric for Meteorol 170:146–165. https://doi.org/10.1016/j.agrformet.2011.10.003

Tso TC (2004) Agriculture of the future. Nature 428:215–217. https://doi.org/10.1038/428215a

Van Vuuren DP, Edmonds J, Kainuma M, Riahi K, Thomson A, Hibbard K, Hurtt GC, Kram T, Krey V, Lamarque JF (2011) The representative concentration pathways: an overview. Clim Change 109:5–31. https://doi.org/10.1007/s10584-011-0148-z

Wang HJ, Sun JQ, Chen HP, Zhu YL, Zhang Y, Jiang DB, Lang XM, Fan K, Yu ET, Yang S (2012) Extreme climate in China: Facts, simulation and projection. Meteorol Z 21:279–304. https://doi.org/10.1127/0941-2948/2012/0330

Wang Y, Liao W, Ding Y, Wang X, Jiang Y, Song X, Lei X (2015) Water resource spatiotemporal pattern evaluation of the upstream Yangtze River corresponding to climate changes. Quat. Int. 380–381:187–196. https://doi.org/10.1016/j.quaint.2015.02.023

Wu K, Wang S, Song W, Zhang J, Wang Y, Liu Q, Yu J, Ye Y, Li S, Chen J, Zhao Y, Wang J, Wu X, Wang M, Zhang Y, Liu B, Wu Y, Harberd NP, Fu X (2020). Enhanced sustainable green revolution yield via nitrogen-responsive chromatin modulation in rice. Science 367(6478):eaaz2046. https://doi.org/10.1126/science.aaz2046

Xiao D, Tao F (2016) Contributions of cultivar shift, management practice and climate change to maize yield in North China Plain in 1981–2009. Int J Biometeorol 60:1111–1122. https://doi.org/10.1007/s00484-015-1104-9

Xiao D, Liu D, Wang B, Feng P, Bai H, Tang J (2020). Climate change impact on yields and water use of wheat and maize in the North China Plain under future climate change scenarios. Agric. Water Manage. 238:106238https://doi.org/10.1016/j.agwat.2020.106238

Xinhua (2015). Potato upgraded as new staple crop. China Daily 2015-01-08. https://www.chinadaily.com.cn/china/2015-01/08/content_19269910.htm

Xu Y, Xu C, Gao X, Luo Y (2009) Projected changes in temperature and precipitation extremes over the Yangtze River Basin of China in the 21st century. Quat Int 208:44–52. https://doi.org/10.1016/j.quaint.2008.12.020

Yang W, Di L, Sun Z (2021) Groundwater variations in the North China Plain: monitoring and modeling under climate change and human activities toward better groundwater sustainability.

Chapter 5 in: Mukherjee A, Scanlon BR, Aureli A, Langan S, Guo H, McKenzie AA (eds), Global Groundwater. Elsevier, pp 65–71

Ye JS (2014) Trend and variability of China's summer precipitation during 1955–2008. Int J Climatol 34:559–566. https://doi.org/10.1002/joc.3705

Zhang XY, Chen SY, Sun HY, Wang YM, Shao LW (2010) Water use efficiency and associated traits in winter wheat cultivars in the North China Plain. Agric Water Manag 97:1117–1125. https://doi.org/10.1016/j.agwat.2009.06.003

Zhang X, Chen S, Sun H, Shao L, Wang Y (2011) Changes in evapotranspiration over irrigated winter wheat and maize in North China Plain over three decades. Agric Water Manage 98:1097–1104. https://doi.org/10.1016/j.agwat.2011.02.003

Zhang HL, Zhao X, Yin XG, Liu SL, Xue JF, Wang M, Pu C, Lal R, Chen F (2015) Challenges and adaptations of farming to climate change in the North China Plain. Clim Change 129:213–224. https://doi.org/10.1007/s10584-015-1337-y

Zou Y, Yang X, Pan Z, Sun X, Fang J, Liao Y (2008) Effect of CO_2 doubling on extreme precipitation in Eastern China. Adv Clim Change Res (Chinese Edition) 4(2):84–91

Final Remarks

The solution of the NCP's water problems is determined by the balance of three big forces: the national requirement of grain security, the water resources system, and the agricultural system. Water authorities are in favor of water transfers while agricultural authorities are in favor of subsidized adaptation of the cropping system. However, there is no silver bullet. The solution of the NCP's water problems until 2050 and beyond consists of a mosaic of measures, which needs all available pieces to complete the puzzle. Some pieces are not easily recognized as contributors. China rightfully relaxed its requirements of grain self-sufficiency in recent years. Imported grain is imported virtual water. Grain imports of 99 Mio. tons in 2020 were at a record high compared to the local production of 670 Mio. tons. They make a non-negligible contribution to the alleviation of water scarcity as do the imports of meat of all kinds. Other pieces of the puzzle are for example irrigation with wastewater—after proper treatment—and seawater desalinization.

Actors should be encouraged to contribute their inventive share, hopefully avoiding the pitfalls existing in every method. They can all be guided by the value of irrigation water expressed in terms of a stiff water price. The government need not make people pay that price in every case. A tax and fee system makes only sense if it leads to a reduction of pumping, if not directly via farmers' profit optimization then by turning the revenue into subsidies to incentivize proven water saving methods in agriculture.

When discussing climate change, we looked at scenarios for 2035–2064 and for 2070–2099 (Sect. 5.5). If we look that far ahead, we also must look at other societal developments. Towards the end of the century, China's population will be half of that of today (see Box 5.2). Then water resources will easily get into equilibrium with demand, and management can concentrate on issues of water quality, for example salinity and pollution, which are probably the more serious long-term problems. For the coming 30 years, however, China still has to sustain or even increase the food production of today to satisfy the demands of her people. That implies that groundwater management in the NCP as outlined in this book will still be needed

W. Kinzelbach et al., *Groundwater Overexploitation in the North China Plain: A path to Sustainability*, Springer Water, https://doi.org/10.1007/978-981-16-5843-3

for a while to see China over the population peak before stress on water resources ceases towards the end of the century.

To conclude, we see the continued need of managing groundwater sustainably, avoiding not only over-pumping but increasingly deterioration of groundwater quality. If suitably managed, groundwater resources are cheap, readily accessible, and more importantly, provide a safeguard to buffer extremes under climate change.

Appendices: Technical Guides and Manuals of Tools

Nine appendices are provided, containing more technical details on methods used, documentation of software and instructions for use and further development. They are listed below, together with the book chapters they refer to. The appendices can be downloaded from Springer's website.

A-1	**Conversion of Pumping Electricity to Groundwater Abstraction** Supplementing Chapter 4.2.3
A-2	**CropMapper** Supplementing Chapter 4.2.4
A-3	**Guantao Box Model (including excel sheet)** Supplementing Chapter 4.3.1
A-4	**Guantao 2D Groundwater Model** Supplementing Chapter 4.3.2
A-5	**Guantao Real-time Groundwater Model** Supplementing Chapter 4.3.3
A-6	**Guantao Data Driven Model** Supplementing Chapter 4.3.4
A-7	**Guantao Irrigation Calculator** Supplementing Chapter 4.4.2 and Boxes 4.2 and 4.3
A-8	**Guantao Decision Support System** Supplementing Chapter 4.4
A-9	**Save the Water—Groundwater Game (including materials for board game)** Supplementing Chapter 3.3

© The Editor(s) (if applicable) and The Author(s) 2022
W. Kinzelbach et al., *Groundwater Overexploitation in the North China Plain: A path to Sustainability*, Springer Water,
https://doi.org/10.1007/978-981-16-5843-3

Printed in the United States
by Baker & Taylor Publisher Services